Biotechnological Interventions for Dryland Agriculture
Opportunities and Constraints

Biotechnological Interventions for Dryland Agriculture
Opportunities and Constraints

Editors

G. Pakki Reddy
Professor and Coordinator

P. S. Janaki Krishna
Subject Expert (Biotechnology)

Andhra Pradesh Netherlands Biotechnology Programme,
Biotechnology Unit, Institute of Public Enterprise, OU Campus,
Hyderabad-500 007, India.

BSP BS Publications
A unit of **BSP Books Pvt., Ltd.**

4-4-309, Giriraj Lane, Sultan Bazar,
Hyderabad - 500 095 - A.P.
Phone : 040 - 23445605, 23445688

IPE
BTU

Published by

BSP **BS Publications**
A unit of **BSP Books Pvt., Ltd.**

4-4-309, Giriraj Lane, Sultan Bazar,
Hyderabad - 500 095 - A.P.
Phone : 040 - 23445605, 23445688
e-mail : info@bspbooks.net
www.bspublications.net

ISBN : 978-93-52300-04-4 (HB)

Foreword

After the information and communication revolution the mankind is witnessing yet another revolution i.e, biorevolution. The discovery of the double helical model of DNA by Watson and Crick in 1953 and later developments in life sciences have made it possible to introduce a number of innovative technologies, which have a profound influence on the nature of products and processes. Biotechnologies emerging from different branches of life sciences have already made deep inroads into health sector and continuing to surge ahead in agriculture and a number of other sectors like animal stock breeding, industrial processing, food processing, nutrition waste treatment etc. The applications of biotechnologies in agriculture said to have far reaching consequences particularly in developing countries in terms of food security and nutritional improvement. A number of industrialized nations share this view and even support the developing countries under bilateral cooperation to augment their resources to undertake research and develop appropriate biotechnologies. The impact of such investments seems to be trickling down with ever-increasing areas under genetically modified crops world over which has already crossed 81 million hectares. However, it is argued that the technologies are supply driven and profit oriented. In most cases, they do not even address the needs of common man in developing societies. Against this backdrop, there are certain attempts to change the course of direction of these promising technologies to face the challenge of providing safe and adequate food in the hunger stricken developing countries. Such initiatives are inspired by liberal donor agencies and local governments in collaboration with civil societies. One such initiative is the Andhra Pradesh Netherlands Biotechnology Programme that promotes demand driven (tailor made) biotechnologies with Dutch public funding, in one of the federal states of India, the second largest populous country on the planet. While developing a number of bioproducts, the Programme over the years established participatory processes for development of biotechnologies relevant to resource poor farmers. I am pleased that this Programme organized an international workshop on "Biotechnological Interventions for Dryland Agriculture : Opportunities and Constraints" and the proceedings are being published in the form of a book. I am glad that the present book focuses on the potential of biotechnology with regard to genetic engineering and molecular markers and other important relating issues like biosafety and human resource development. I am impressed immensely by the range of topics that are covered in this book. Finally, I complement the editors for the excellent compilation of papers presented in the book which I certainly hope will be very useful for all stakeholders involved in the promotion of biotechnology.

M. V. Rao

Chairman, Biotechnology Programme Committee,
Andhra Pradesh Netherlands Biotechnology Programme,
President, Indian Society of Genetics and Plant Breeding,
Former Special Secretary, DARE,
Former Special Director General, ICAR,
Former Vice Chancellor, ANGRAU,
India.

Preface

We are happy to place this book before you after a regrettable delay. The present book includes the proceedings of the International Workshop on "Biotechnological Interventions for Dryland Agriculture: Opportunities and Constraints" organized by the Andhra Pradesh Netherlands Biotechnology Programme, Institute of Public Enterprise, Hyderabad, India during 18th – 20th July 2002 at Hyderabad.

Agricultural biotechnology is one of the most rapidly growing technologies in the history of agriculture known to mankind. Globally the area under genetically modified/biotech crops increased almost 45 fold since their introduction in 1996. This is the highest growth rate of any technology in agriculture sector. Products of biotechnology are being commercialized and in particular are now being traded in the international markets. However, crops and traits focused so far are of commercial importance and the crops grown in drylands are yet to benefit from biotechnological interventions. Improving both the productivity and sustainability of rainfed agriculture in the more difficult marginal environments is a challenging task. Arid and semi arid regions comprise almost 40% of the world's land area. Approximately two-thirds of these drylands are in developing countries. Also growth of rainfed agriculture is nearly stagnant in arid and semi arid tropics, and green revolution technologies have had little impact in these areas. The challenge, therefore, is to introduce technologies that increase agricultural production and sustainability in these complex, diverse and risk prone areas. Biotechnology appears to be one of the options to meet this challenge. The present international workshop was designed to address some of the key issues in dryland agriculture and the potential of biotechnology in tackling these issues.

The major topics discussed in the workshop related to the status and potential of biotechnology with regard to dryland crops. Particularly the techniques of molecular markers and genetic engineering are elaborated. Human resource development, biosafety and risk assessment with regard to biotechnology are also discussed. Presentations and discussions indicated that enhancement of productivity in food grains, legumes and oilseed crops should become an important research goal. Participants identified the need for improved understanding of the molecular basis of pathogencity and designing tightly regulated pathogen-specific promoters. Abiotic stress tolerance research should be focused on the level and regulation of genes expressed during drought. Also, examining the role of heat shock proteins and transcription factors in regulating various biosynthetic pathways under stress conditions could lead to the development of drought tolerant genotypes. The presentations on molecular markers detailed their potential in crop breeding, genomics and assessing genetic diversity and germplasm conservation.

Speakers emphasized that molecular markers are important tools for a large number of applications, ranging from locating genes to improving plant varieties. The presentations on critical issues pertaining to applications of biotechnology focused mainly on biosafety and risk assessment with regard to genetically modified crops and foods. The success of biotechnology invariably depends on trained and skilled manpower. The presentations on training and educational needs provided an overview of manpower and human resource requirements from the public institutions and industry perspective also. The importance of university and industry collaborations in training students was recognized, as was the need to regulate the quality of education offered in modern biology courses. Invited speakers addressed various issues in their formal presentations and have been included in this book in the form of papers. Some speakers provided only 'Notes' on their presentations hence we are constrained to include their full papers. Nevertheless, we assure that the issues raised and covered in this book are current as they have been all these years and they continue to be so for sometime to come.

In this endeavour, we are thankful to all the persons involved in the workshop either directly or indirectly as organizing an international event is a challenging task. Our thanks are due to the Ministry of Foreign Affairs, the Government of the Netherlands for the financial support. We are deeply indebted to Dr C. Rangarajan, the then Governor of Andhra Pradesh, India for inaugurating the workshop and Dr G.S Khush, International Rice Research Institute, Philippines for giving an informative Key Note Address. Our thanks are due to Mr Theo van de Sande, for his valuable 'Message' and his active participation during the workshop. We are grateful to Dr M V Rao, Chairman, APNLBP for his constant guidance, encouragement and support during the workshop. Our thanks are due to Dr. Mangala Rai, the then DDG, Crop Sciences, ICAR, Indian Council for Agricultural Research, New Delhi for his thought provoking, inspiring valedictory address. Our acknowledgements are due to all the Members of the Andhra Pradesh Netherlands Biotechnology Programme Committee for their constant support. We are indeed grateful to all the speakers and participants for their contributions and for their enormous patience in waiting for this publication to see the light of the day.

Please note that although the Ministry of Foreign Affairs, Government of the Netherlands supported this workshop as part of the Programme activities the opinions expressed in the proceedings of the workshop do not represent the official policy or the position of the agency. All the opinions expressed in the book strictly belong to the authors and individuals who participated in the workshop. It is our hope that this book will provide useful insights to the readers with regard to the status of biotechnologies in dryland agriculture and provide a direction to develop appropriate technologies for the benefit of dryland farmers.

Editors

Sept. 2005
Hyderabad, India.

G Pakki Reddy
P S Janaki Krishna

Contents

Biological Interventions and Dryland Agriculture

Dr. C. Rangarajan
His Excellency. The Governor of Andhra Pradesh.

I am happy to be here in your midst this morning to inaugurate the International Workshop on "Biological Interventions and Dry Land Agriculture : Opportunities and Constraints". Needless to say, the theme you have chosen is appropriate and timely.

A major challenge before the world during the 21st century will be to ensure food security for all. While Malthus has been proved wrong, the issue he raised still haunts the world. From time to time, fears have been expressed whether the food sources would be sufficient to feed an ever increasing world population. An estimated increase of more than 90 million people a year will be the largest in human history. While we in India claim self-sufficiency in production in food grains, two thirds of Indian children under age five are malnourished. Let us recall that the Food and Agricultural Organisation of United Nations defines "food security" as a *"state of affairs where all people at all times have access to safe and nutritious food to maintain a healthy and active life"*. India has still several miles to go to reach this goal. It is against the background of continuing population growth, accelerating urbanisation and high level of poverty and the consequent pressure on the social system that the question of whether food security can be achieved in future is posed.

Challenges Before Indian Agriculture

Agriculture in India, as in many developing countries, continues to play a critical role in the country's overall economic growth and development. Despite some significant shifts in the structure of the Indian economy, even today, two thirds of the population are dependent on agriculture. To absorb the growing labour force a strong growth in agriculture becomes almost imperative. During the 60s, far reaching decisions were taken by the Government to modernise agriculture to meet the challenges posed by the burgeoning population. Over the period from 1960 to date, yield levels have increased by three times in rice and six times in wheat. These yield increases are attributed largely to improved varieties of seeds, irrigation, fertilisers and a range of improved resource management practices. This has been the underlying strategy of "Green Revolution". In a sense, not only India but several parts of the world were swept by this revolution at that time. While "Green Revolution" in India ushered in an era of food self-sufficiently, these productivity increases have also contributed to environmental problems, such as increase in soil salinity and lower water tables in irrigated areas, human health problems due to excessive pesticide use and water pollution and soil degradation resulting from excessive use of chemical inputs. A recent study shows that if the desired growth rate in agriculture of four per cent per annum is to be achieved through the currently accepted technology mix, the per hectare chemical fertiliser consumption will have to increase to 185.81 kgs in 2011 - 2012 from the current level of 75.26 kgs per hectare. It is obvious that such a level of chemical fertiliser intensity would be undesirable for both environmental and long run productivity reasons. Can bio-technology be an answer to the problem? Can it result in enhancing agricultural growth without causing serious environmental problems?

"Green Revolution" technologies by their very nature have had very little role to play in enhancing agricultural growth in rain-fed areas. However, dry land agriculture continues to play a very important role in the total food and agricultural production. Dry land agriculture in India covers 67 per cent of the net cultivated area and currently accounts for more than 60 per cent of the food grains, –90 - 95 per cent of millets, almost 80 per cent of oil seeds, 90 per cent of the green legumes and 70 per cent of cotton. Even 50 per cent of the paddy is grown under rain-fed conditions. Increasing the productivity of dry land agriculture becomes absolutely essential also for the reason that only the improvement in the productivity of dry land agriculture can lead to benefits of growth being shared by many. Improving the productivity and ensuring the sustainability of dry land agriculture are, however, a challenging task.

Biotechnology and its Implications

As we cannot add to the area of arable land, measures to wrest more food from the land we have with less energy input have become urgent. In this context, amongst several technological options, biotechnology is emerging as a powerful possibility. Biotechnology may be broadly defined as technology that deals with living organisms. Modern biotechnology can help better plant breeding, because of the breakthroughs achieved in molecular and cellular biology by way of gene identification and its manipulations. Biotechnology can enhance product quality by improving the characteristics of plants and animals. It may also potentially conserve natural resources and improve environmental quality by using organisms for degradation of toxic chemicals and wastes, as fertilisers for soil improvement and the development of insect and disease resistant plant varieties. Modern biotechnology can thus help in better plant breeding, plant production, product quality and conservation of natural resources. Norman Borlaug along with a coalition of prominent scientists in a Declaration on 30th April, 2002 said *"Additional high yield practices, based on advances in biology, ecology, chemistry and technology are critically needed in agriculture and forestry not only to achieve the goal of improving the human condition for all people but also the simultaneous natural preservation of the natural environment and its bio-diversity through conservation of the wild areas and natural habitat"*. This Declaration underscores the importance of modern biology for agricultural development.

In determining whether biotechnology is the appropriate new technological paradigm in the fight against hunger, a close look at the perceived benefits and risks becomes necessary. Over the years biotechnology has made concrete advances in the areas of fermentation, biofertilisers, biocontrol agents, tissue culture and genetic engineering. Despite the controversy surrounding transgenic seeds, the area under genetically modified crops is estimated at 52.6 m.ha. as of 2001. Four countries account for 99 per cent of the global transgenic area, the USA occupying the top position with 68 per cent of the total, followed by Argentina (22 per cent), Canada (6 per cent) and China (3 per cent). The principal GM crops grown are soybean (63 per cent), followed by GM corn (19 per cent), transgenic cotton (13 per cent) and GM Canola (5 per cent).

Though the tools of biotechnology developed over the last thirty years have clearly opened up dramatic opportunities to create new varieties of plants and animals, the novelty of biotechnology has raised several issues, the most important of which relates to biosafety. Some view biotechnology as a logical and modest extension of conventional plant and animal breeding technologies; others see it as a novel technology that is entirely different from traditional plant breeding. Some fear that it can lead to 'biodevastation'. Some talk of 'Frankenstein foods'. How different is biotechnology from traditional plant and animal breeding? What kind of problems is it attempting to solve? What are the future uses of biotechnology? What are the principles on which biotechnology should be promoted ? What are the enabling conditions we need to create for rapid growth of this technology? These are some of the important questions that need the attention of an workshop like this.

Need for Consensus

It is high time that scientists come out with a clear assessment of the role of genetically modified crops. Every technological advance brings with it potential benefits and risks. However, the scientific community has an obligation to state whether on balance the introduction of transgenic crops is good for the society or not. Continuing controversies on such issues are harmful. India is one of the few countries where a regulatory system with respect to biosafety is in place. Despite this, controversy continues, for example, the permission given for the introduction of Bt cotton. It is in this context that this Workshop is extremely important. You must address this issue squarely and come out with your views. While the country and the farmers should not be exposed to unnecessary risks, we must also take care to ensure that we do not miss out on the benefits of a new technological advance. The community will be guided by the consensus or near consensus among the scientists and experts.

Broadly speaking, for the successful promotion of biotechnologies two basic principles appear to be necessary to be followed - the principle of precaution and the principle of transparency. In view of the implications of biotechnology on human and animal health and environment utmost care need to be taken before the release of these technologies. They should be subjected to elaborate risk assessment to find out all possible effects both intentional and unintentional, positive and harmful. However, the precautionary principle should not prevent reaping the advantages of the new technology. In the same way, the process of decision making permitting the use of biotechnologies should be transparent. This means that all the information relevant to the issue of permits should be made public.

As of now, the agricultural biotechnology is dominated by big private companies in industrialised countries. This raises suspicion in the minds of some people that the technologies might be misused to the disadvantage of the already vulnerable sections in the developing countries. It is possible that such situation could be averted by strengthening the capacities of developing countries to develop biotechnologies appropriate to their local needs. In this context, investments in public sector R & D need to be augmented. This is all the more necessary because there is a declining trend in public investment in R & D sector. Enhanced public investments in biotechnology in developing countries will not only improve public confidence but also help in overcoming a number of legal complications associated with Intellectual Property Rights.

Dryland Agriculture

The problems associated with dryland agriculture are more complex as compared to irrigated agriculture. The most important issue with regard to agriculture in dry land is conservation of water which is a scarce commodity in these areas. Increasing yields in dryland agriculture involves identifying existing crop varieties or breeding new ones which have better water use efficiency and can come to harvest within a short period of time. The need of the hour is to introduce the capacity to tolerate water stress and resistance to pests and diseases in the crops cultivated in these areas. Sorghum, groundnut, castor, minor millets, vegetables like tomato need special attention. Before considering the adoption of high-tech technologies such as biotechnology it should be emphasised that the most important consideration is the capability of the agroecological situation prevailing in the dry land ecosystem to use the technology. For example, if a genetically modified crop variety such as groundnut is cultivated under sub-optimal conditions, the yield of the newly introduced variety will remain well below its potential. The low end of biotechnological applications such as tissue culture, cultivation of mushrooms, vermicompost, biofertilisers and biopesticides are cost effective and therefore suitable for adoption in dry land situations. The work carried out by the Andhra Pradesh Netherlands Biotechnology Programme in introducing such technologies through NGOs and research institutions has shown encouraging results. Application of high end biotechnological tools such as genetically modified crops may also have an important role subject of course to meeting the biosafety standards. We need to mount separate technology missions with respect to different crops grown in dryland areas in order to achieve increased productivity.

I am indeed happy that this Workshop is being held here in Hyderabad and in the State of Andhra Pradesh which has taken a lead in promoting the use of biotechnology. Andhra Pradesh is one of the few States which have formulated a distinct policy on biotechnology. A Biotech Park with modern facilities is coming up near Hyderabad. This Workshop is taking place at a critical time when some important decisions in relation to the use of the bio-technology are in the process of being taken. We have the participation of a galaxy of distinguished and eminent scientists including Dr. Gurudev S. Khush. I am sure that the deliberations of this workshop would provide the much needed direction to the use of biotechnology in improving the yields in agriculture in general and in dryland areas in particular. Your deliberations can resolve many controversial aspects relating to the introduction of biotechnology. When I wish your deliberations all success, I do not say this in a formal and conventional sense. The success of your deliberations can lay the foundation for a sound policy.

Food Security and Poverty Alleviation Through Application of Biotechnology

G. S. Khush

Principal Plant Breeder & Head-Division of Plant Breeding, Genetics & Biochemistry International Rice Research Institute, Philippines g.khush@cgiar.org

Access to food, shelter, and safe drinking water is the basic right of every individual born on this planet. Yet billions of people struggle for access to adequate levels of food. They are able to improve their prospects for a better future only when abundant and affordable food is available. Mankind has faced the problems of food insecurity since times immemorial. There are numerous reports of food shortages and famines during recorded history. As late as 1940s in India, 1950s in China, and 1970s in Bangladesh millions of people perished due to famines.

Three global revolutions have impacted our ability to grow food. First of course was the agricultural revolution starting about ten thousand years ago when mankind became food growers rather than food gatherers. This settled people in small communities and launched the civilization. Our ancestors established the foundations of organized society and domesticated plants and animals resulting in several-fold increase in food production. The second global revolution, the industrial revolution was the harbinger of enormous change in production methods. The process of production and specialization led to enormous burst of output, bringing big improvements for much of humanity during the last two centuries. It impacted the food production immensely particularly in the developed world through mechanization. The third global revolution, e.g. the revolution in information technology or (IT) and biotechnology, (BT) is impacting the food production in several ways. I shall come to these technologies later in my talk.

The 1960s was a decade of despair with regard to the world's ability to cope with food-population balance, particularly in the developing countries. The cultivated land frontier was closing in most Asian countries, while population growth rates were accelerating, owing to rapidly declining mortality rates resulting from advancements in modern medicine and health care. International organizations and concerned professionals were busy organizing seminars and conferences to create awareness regarding the ensuing food crisis and to mobilize global resources to tackle the problem on emergency basis. In a famous book entitled "Times of Famine" published in 1967, the Paddock brothers predicted and I quote, "Ten years from now, parts of the underdeveloped world will be suffering from famine. In 15 years, the famine will be catastrophic, and revolutions and social turmoil and economic upheavals will sweep areas of Asia, Africa, and Latin America."

Thanks to widespread adoption of green revolution technology, large-scale famines and social and economic upheavals were averted. Between 1966 and 2000 world population grew by 90% but food production increased by 130%. In 2000, the average per capita food grain availability was 20% higher than in 1966. The engine driving the Green Revolution was the high yielding, disease and insect resistant and fertilizer responsive varieties of rice, wheat, and maize. The adoption of these varities was facilitated through expansion in the irrigated area, the availability of inorganic fertilizers and benign public policies. Proper water management and control is the key to the adoption of green revolution technology. There was tremendous investment in the development of irrigation facilities during the 1960s and 1970s. The irrigated land in the world increased from 94 million hectares in 1950 to 240 million hectares in 1990, an increase of 2.4% per year. In 1997 China and India had 50 and 55 million hectares of irrigated land respectively. Fertilizer use contributed to increased productivity. Worldwide fertilizer use increased from 14 million tons in 1950 to 140 million tons in 1990, a tenfold increase. The adoption of green revolution technology was facilitated by government investment in infrastructure such as roads, markets, electrification and price support for inputs and produce.

Gradual replacement of traditional varieties of cereals by improved ones, together with associated improvements in farm management practices has had a dramatic impact on the growth of food production. Farmers harvest 5-7 tons of unmilled rice from high yielding varieties as compared to 1-3 tons from conventional varieties. Since 1966, when the first high yielding variety was released, the total rice production more than doubled from 257 million tons in 1966 to 600 million tons in 2000. Similarly world wheat production increased from 308 million tons in 1966 to 590 million tons in 2000. In Asia, wheat production increased from 33 million tons to 225 million tons in 1995 or a sixfold increase in a 30 year period. Rice production increased threefold in India, the Philippines and Thailand, fourfold in Indonesia and fivefold in Vietnam. In most of the rice growing countries the growth in rice production has outstripped the rise in population leading to a substantial increase in consumption and calorie intake. The increase in per capita availability of rice and wheat and decline in the cost of production per ton of output contributed to a decline in real price of rice and wheat in international and domestic markets. The unit cost of production is about 20-30% lower for high yielding varieties than for traditional varieties of rice and wheat. Price of rice and wheat adjusted for inflation is now 50% lower than in mid-1960s. The decline in food prices has benefited the urban poor and rural landless who spend 50-60% of their income on purchasing food.

High yielding varieties require more labor per unit of land because of intensive care in agricultural operations and harvesting of a larger output. The labor requirement has also increased because of the higher intensity of cropping which has been made possible by the reduction in duration of crop growth. The marketing of a larger volume of produce and an increased demand for non-farm goods and services, resulting from larger farm incomes has generated additional employment in rural trade, transport and construction activities. The rapid economic development in many Asian countries including Thailand was triggered by the growth in agricultural income and its equitable distribution which helped expand the domestic market for non-farm goods and services. Availability of cheap food and rising living standards have contributed to political stability.

Increases in food grain output has resulted in environmental sustainability. More than half of the world's poor live on lands that are environmentally fragile and rely on natural resources over which they have little control. Land hungry farmers resort to cultivating unsuitable areas such as erosion-prone hillsides, semi-arid areas where soil degradation is rapid and tropical forests where crop yields on cleared land drop sharply just after a few years. The availability of crop varieties with multiple resistance to diseases and insects reduced the need for application of agrochemicals and facilitated the adoption of integrated pest management practices. Reduced insecticide use helps (i) enhance environmental quality (ii) improved health of farming communities (iii) makes safer food more available (iv) protects useful fauna and flora.

The Food Grain Situation at the Beginning of the 21st Century

In spite of the achievements of green revolution 800 million people in the world go to bed hungry everyday. Alarm bells are ringing again about our ability to meet the food grain requirements in the new millennium. We must ponder over the situation.

Some of the factors leading to this scenario are :

◆ *Declining rate of growth of food production*

From 1960 to 1990, the world grain harvest increased from 847 million tons to 1780 million tons. This was a remarkable achievement in the history of agriculture. Cereal grain production expanded at the rate of almost 3.0% during 1970s and 1980s. Since 1990 however, the growth in grain harvest has slowed to about 1.2%. We must develop the technologies to enhance the rate of food grain production.

◆ *Increasing population in developing countries*

World population was 1 billion in 1850. It took 80 years to reach 2 billion in 1930, 30 years to reach 3 billion in 1960, fifteen years to reach 4 billion in 1975 and 12 years to reach 5 billion in 1987 and 13 years to reach 6 billion in 2000. From a high of 2.2 % in 1964, the population growth rate has declined to 1.4% per year now. Most of this increase is occurring in the developing countries. The population has stabilized in 32 countries mostly in Europe and Japan; in four countries the population growth rate is actually negative. In contrast to this group, some developing countries are projected to triple their populations by 2050. For example Ethiopia's current population of 62 million will more than triple. Pakistan's population is growing at the rate of 2.8% per year and is projected to grow from 148 million to 357 million, surpassing that of the USA before 2050.

The population of Nigeria, meanwhile, is projected to grow from 122 million to 339 million, giving it a higher population in 2050 than for the whole of Africa in 1950. Per capita land availability has declined from 0.5 hectares per person in 1960 to 0.27 hectares per person in 2000. It will be 0.1 hectare per person in 2030. Under UNDP predictions, the world population is likely to reach 8 billion in 2030.

◆ *World poverty*

1.3 billion people in the world are absolutely poor, somehow surviving on an income of less than one dollar a day. Another two billion are marginally better off. In 47 least developed countries of the world, 10% of the world's population subsists on less than 0.5% of the world's income. Some 40,000 people die from hunger-related causes everyday. The top 20% of the world's population consumes 85% of the world's income, remaining 80% live on 15%. The three richest persons on the planet have more wealth than the combined GDP of 47 poorest countries. The richest 15 persons have more wealth than the combined GDP of all of Sub-Saharan Africa with its 550 million people.

Poverty, agriculture and environmental problems are interrelated. Poor people live in fragile environments, practicing subsistence farming leading to environmental problems and even lower productivity. This subsistence agriculture is the cause and result of poverty. If and when poverty alleviation programs succeed, the purchasing power of the poor will improve and so will the demand for food.

◆ *Changing food habits*

The most important factor that influences per capita consumption of staple grains is the level of income of the consumer. At low levels of income, meeting energy needs is the most basic concern of an individual. Staple foods such as starchy roots, rice, wheat and coarse grains provide the cheapest source of energy. Low income consumers spend most of their income on these types of food. As income increases, the consumer shifts from low-quality to better quality products. For example, rice is the most preferred food staple in Asia where 90% of the world's rice is produced. At low levels of income, rice is considered a luxury. However, as income increases beyond a threshold level, consumers can afford to have a high-value balanced diet containing foods that provide more proteins and vitamins, such as vegetables, fruits, eggs, milk, and meat. This pattern of change in food consumption with economic growth is amply demonstrated by the experience of Japan and Korea which made a transition from low to high income level within a short period of time. The rice consumption in Japan increased with economic growth after the Second World War, reaching a peak of about 120 kgs per capita per year in early 1960s and then started to decline. By the late 1980s the per capita rice consumption was 40% lower and it is only 68 kg per person per year.now. South Korea followed a similar pattern albeit somewhat later than Japan. The FAO data show that per capita cereal consumption has started to decline in mid-income countries such as Singapore, Malaysia, and Thailand. China and Indonesia are reaching the threshold of peak cereal consumption.

In projecting the growth in demand for cereals, we must consider their indirect demand as livestock feed. Asia as a whole has emerged as a major consumer of livestock products. It takes 2,4, and 8 kgs of grains to produce 1 kg of poultry, pork, and beef respectively. This increase in demand for livestock products implies a fast growth in demand for cereal grains as livestock feed.

◆ *Projections for food grain demand*

Keeping the above factors in mind, FAO statistics show that food grain production must increase from 2.1 billion tons now to 3.1 billion tons in 2030 or an increase of 50%. It should increase by 5% in Asia, 100% in Latin America, and 400% in Africa. That is a big challenge for scientists, government policy makers and farmers.

Technologies for Increasing the Food Production

Tabulation for meeting the challenge of achieving food security we must develop crop varieties with higher yield potential and greater yield stability and more efficient management practices. Conventional plant breeding methodologies as well as emergent frontier technologies must be exploited to develop crop varieties for the 21st century. Time tested methods of crop improvement such as hybridization and selection, hybrid varieties and modification of plant architectures will continue to be employed. Hybrid varieties of rice for example yield 15% more than the true breeding varieties. New plant type varieties popularly called Super Rice developed at IRRI yield 20% more than the widely grown high yielding varieties.

Amongst the frontier technologies for crop improvement, molecular marker aided selection and genetic engineering have captured the imagination of crop scientists and policy makers alike. Construction of dense molecular genetic maps of major food crops has ushered in an era of molecular markers which are being employed in moving genes from one varietal background to another and for pyramiding or combining of several genes for the same trait such as disease or insect resistance, through molecular marker aided selection. This technology is being exploited to develop varieties with more durable resistance.

Genetic engineering or recombinant DNA technology is being exploited to introduce cloned genes from unrelated sources into crop varieties for increasing yield, disease and insect resistance and novel grain quality traits. Transgenic crops with value added traits were planted in 42 million hectares in 2001. Farmers reap more profits through increased yield and reduced insecticide costs. Wherever such technology has a role in increasing food production it should be employed. However, this innovative method of crop improvement has generated a lot of debate between its proponents and opponents about the benefits and risks of using genetic engineering technology. If in the interest of national well being this technology has a role, then it is prudent to explore ways by which its advantage can be harnessed without taking undue risks. This assessment of risks and benefits should be made on the basis of scientific criteria in a transparent manner involving various stakeholders. In order that farmers are not deprived of the benefits of improved materials generated through advanced technologies and the enthusiasm of technology developers and practitioners is not suppressed, decisions about commercialization of the product should be taken expeditiously and in a hassle-free procedural environment.

A study at North Carolina State University in USA related to principles for socially responsible biotechnology made the following points.

◆ It is socially responsible to allow and encourage development of new technology products where important benefits are available to society e.g. to combat hunger and protect the environment.

◆ It is socially responsible to take every precaution to ensure that the products of biotechnology are safe for people and the environment. For example, genetically modified crops have undergone extensive tests and have been shown to be safe for humans and the environment.

◆ It is socially responsible for scientists and companies to follow all Government regulations and not try to circumvent them for market gains. For example, no agricultural product should be commercially marketed until it has received full approval for human consumption.

◆ It is socially responsible for credible third-parties to provide interested consumers with as much factual and balanced information as they desire without the imposition of higher costs on consumers.

◆ It is socially responsible for Government agencies to regulate biotechnology based on best available science rather than on politics.

◆ It is not socially responsible to demand "zero risk" from any technology, to hold biotechnology to unreasonable high standards compared to other food production methods.

◆ It is not socially responsible for opponents of biotechnology to scare farmers and consumers or threaten food companies to promote own business interests and political agenda. For example, after the scientific community and Government agencies conclude that there are no adverse health impacts from a biotechnology product, opponents should not continue raising fears to further their own agenda.

This brings me to the promise of genomics era with far reaching implications for human health, food production and human nutrition. The milestone publication of two draft genome sequences of rice, in 2002 brings the cereal crop of the world's poor to the center stage of genomics. These drafts will be combined with a complete rice genome sequence being compiled by the public International Rice Genome Sequencing Project (IRGSP). The IRGSP sequence is expected to be published later this year. The highly accurate IRGSP sequence will serve as a gold standard for all future investigations of genetic variation in crops. Rice, the world's most important cereal crop for human consumption, is the staple food of more than three billion people, many of them desperately poor. The sequencing of rice genome will benefit many other species particularly cereals such as wheat, maize and sorghum as the gene content and gene order in these cereal crops are similar.

Comparative genomic analysis enables biologists to assign a tentative gene function to a gene according to what that gene does in another species. The availability of rice genome sequence will now permit identification of the function of each one of approximately 50,000 rice genes through functional genomics. Once the function of a gene is verified, new plant varieties can be developed by introduction of the gene through traditional breeding in combination with marker assisted selection or direct engineering of the gene into rice or other cereals. Finally knowing the sequence of specific genes will allow us to tap into the natural genetic variation of crop species. There are over 100,000 accessories of traditional rice varieties collected from a broad range of geoclimates. These seeds serve as a pool of natural variants. To date, this wealth of germplasm has remained largely untapped owing to the difficulty of identifying agronomically important genes. Now if a gene has been proven to contribute to a trait of agronomic importance, alleles or alternate forms of this gene can be examined from multiple varieties for their relative usefulness. Applying the information to food production will require integrated approaches using diverse germplasm, traditional breeding, modern technologies and emerging knowledge from comparative genomics.

The progress in rice improvement during the Green Revolution era discussed earlier was the result of international collaboration. A huge number of rice varieties and improved breeding lines were exchanged between different countries each year through an International Network for Genetic Evaluation of Rice, (INGER), coordinated by IRRI. Improved varieties developed in one country were released in another country for on-farm production or to be used as parents in breeding programs of other countries. The massive exchange of germplasm freely was the basis of advances in varietal development. These efforts have improved food security and touched lives of millions of people in the developing world.

Rice is now a genetic model of crop plants and at the frontiers of genomic research. Uptill now most of the rice improvement was carried out by the public or Government-supported institutions and International Rice Research Institute. Improved genetic materials and genetic knowledge was in public domain. However, with advances in genomic research the private sector is venturing into knowledge generation as shown by Monsanto and Syngenta's investment in rice genome sequencing and germplasm improvement to reap the benefits of research. On the one hand, the private investment can bring about new innovations and technologies for farmers. On the other hand, a shift in the balance of public and private investment in rice research has also aroused concerns that some proprietary technologies might become unavailable to those who cannot afford them. Such concerns must be addressed because gene identification, validation and applications are occurring at an accelerating pace. The key question is, can the model of free access to genes, germplasm, technologies and knowledge exist and contribute under increasingly protective environment that exercises intellectual property rights?

The benefits of private investment in crop research are obvious and collaboration between public and private sector should be encouraged. However, in forging such collaboration, the expectations of both communities must be recognized. In case of private sector, the ability to obtain patents on genes or plant products is crucial. Almost all private sector plant breeding depends on proprietary position that permits recovery of research costs. The classic example is hybrid maize in the USA for which biological proprietary protection is derived from the farmer's inability to use the harvested crop as seed. This has encouraged private sector breeding that has increased yields by more than fivefold since 1930. As developing country seed markets are becoming significant to the multinational biotechnology community, some patents are likely to be sought in larger developing nations.

The public sector needs access to the new technology for further research innovation and use in non-commercial environments. The challenge is to develop a shared vision for rice research that will provide the public sector access and freedom to use modern tools and sufficient incentives for private sector to innovate, develop and deliver new rice technologies. In the human genome project, ten pharmaceutical companies and Welcome Trust have agreed to fund and create a publicly available archive of human genetic variation. A similar pattern of collaboration is needed in rice functional genomics.

The demand for diverse genetic resources and expertise in functional genomics requires a common platform where information and genetic resources can be broadly shared to accelerate trait discovery. The key elements of this collaborative platform are public access to sequence data, abundant genetic resources for functional assignment, capacity for biological evaluation and incorporation of genes and traits into varieties. The arrangements for the release of rice sequence data and donation of golden rice are encouraging signs that private sector is willing to share their products and knowledge freely to developing countries.

Different processes, resources and expertise are needed to reach the final objective of rice improvement. Diverse genetic resources held by rice growing countries and public institutions are crucial to success and these include mutants, germplasm, near isogenic lines, populations for mapping and breeding lines. Furthermore, a tight coupling of the processes involved in assigning gene function and generating improved varieties is essential to ensure that discoveries are put to practical use. Therefore IRRI proposed the formation of an international working group on functional genomics. It was agreed that the following activities are of high priority: (i) create an information node to deposit and disseminate information on rice functional genomics; (ii) build a public platform to promote access to genetic stocks and phenotypic information; (iii) develop databases on phenotypes and mutants with linkages to sequencing laboratories and (iv) initiate partnership to develop resources for microarray analysis.

The pattern of rights envisioned is that genetic resources for functional genomics will be made available to the public and private sector under a material transfer agreement (MTA). This agreement permits recipients to obtain patents and genes discovered through the use of material, but requires them to make available rights under those patents at a reasonable royalty for application in commercial markets of the developing world and at zero royalty for application in noncommercial subsistence farming. In addition to ensuring the possibility of use in the developing world, it is also essential that data and materials are freely available for research. Hence, the envisioned MTA will have provisions permitting free use for research purpose of any patents, as well as provisions ensuring that recipients cannot obtain any form of intellectual property protection on the stocks per se. I am happy to note that India is a member of the Working Group on Functional Genomics.

The International Functional Genomics Working Group has prepared an action plan :

◆ *Information node*

A website has been created where progress on rice functional genomics can be communicated. It will serve as the entry point for funding and sharing information and provide a link to individual laboratories or organizations. The site will also serve as a clearing house of information on genetic resources and their availability.

◆ *Genetic resources*

IRRI has begun to share genetic stocks through MTA that will support the common goals. This includes genetic materials – mutants, isogenic lines and mapping populations – specifically created for gene identification.

◆ *Microarray resources*

A nucleus of laboratories has agreed to contribute expertise and resources essential for making microarrays. A microarray facility has been set up at IRRI which is being used for research and training of scientists for collaborating countries.

◆ *Bioinformatics*

One of the important components of functional genomics program is strong bioinformatics capacity. A bioinformatics unit has been established at IRRI. Collaborating national programs are advised to strengthen bioinformatics capacity.

Finally it is obvious that for reaping the benefits of breakthroughs in biotechnology including genomics and information technology, developing countries must move towards knowledge-based society. The connectivity and the preparation of human resources and their deployment will be the key to enabling developing countries to improve their situation. There is a vast and growing gap in the training of scientists and engineers between wealthy developed countries and the developing countries. For example the United States and Japan have about 70 researchers and engineers per 10,000 population, EU countries about 40, and China and India about 6. The poorest developing countries in Africa have less than one. This gap must be narrowed.

3

Agricultural Biotechnologies for Resource Poor Farmers : A case Study of the Andhra Pradesh Netherlands Biotechnology Programme

G. Pakki Reddy and P.S. Janaki Krishna

Andhra Pradesh Netherlands Biotechnology Programme, Institute of Public Enterprises, Hyderabad, India - 500 007

The Andhra Pradesh Netherlands Biotechnology Programme (APNLBP) is one of the four country programmes supported by the Ministry of Foreign Affairs, Government of the Netherlands. The purpose of the Programme is to strengthen the capacities of local institutions to develop tailor made biotechnologies based on the felt needs of local communities. The unique feature of the Programme is that decisions on research priorities and funding, implementation and monitoring of projects are made by a multi stakeholder steering committee consisting of NGOs, researchers and policy makers. The research agenda of the Programme vastly differs from that of big multinational companies whose interests bypass the concerns of small scale producers. This paper provides a brief description of the process through which the Programme has been striving to promote application of biotechnologies, both traditional and modern, for the welfare of resource poor farmers in rainfed ecosystem.

The Genesis

The substantive phase of the APNL Biotechnology Programme started from 1996 after two years of an elaborate preparatory phase. The initial duration of the Programme was for a period of six years with an outlay of Rs.16.00 crores, which has since been extended upto 2007 with an additional budget of Rs 27.52 crores.

Objectives

⇒ To promote application of biotechnologies relevant to small-scale agricultural producers and processors in Andhra Pradesh, one of the federal states of India.

⇒ To develop appropriate biotechnologies through research activities that focus on identified priority problems;

⇒ To encourage and supplement supportive activities required to ensure development and adoption of biotechnologies including training, transfer of technology, workshops and information dissemination;

⇒ To strengthen capacities of local organizations in the state of Andhra Pradesh to develop and promote adoption of biotechnologies and conduct analysis in the field of technology assessment; and

⇒ To ensure the adoption of biosafety measures and to contribute to discussions on issues of intellectual property and biosafety where appropriate.

Methodology and Approach

The *Interactive Bottom Up* (IBU) approach followed in the APNL Biotechnology Programme is developed basically on the principles of participatory technology development (PTD). The approach emphasises the needs of the farmer/end-user and facilitates the exchange of information amongst all groups, which are involved in the development and application of innovative biotechnologies eventually leading to development of easy to adopt technologies by small and marginal farmers. Following identification of priorities through need assessment surveys the project formulation and implementation was initiated. Although building consensus among multi-stakeholder groups takes time and effort, the experience has been useful as it provides valuable information resulting in a sharper focus on problem areas, a genuine dialogue between users, researchers and policy makers, leading to a consensus on Programme Development.

Public Priority Setting, Planning and Programme Formulation

Followed by an initial preparatory phase and local need assessment survey, a priority setting and planning workshop was organized to bring together all the stakeholders in the Programme. The deliberations of the workshop helped in prioritizing the following specific problem areas in dryland agriculture.

Priorities

(i) Foodgrains and pulses

(ii) Oil seeds,

(iii) Agro-forestry, tree crops, horticulture and sericulture and

(iv) Animal health and production.

Geographical Coverage

The Programme mainly focuses on few selected villages in Mahaboobnagar and Nalgonda districts of Andhra Pradesh. The technologies developed through this Programme are being field tested initially in these villages and their impact assessed. The proven technologies would then be propagated in the other rainfed areas.

Institutional Set-Up

The Programme is unique in the sense that the entire decision making process is vested with local institutions. The donor agency has transferred the ownership to a multistakeholder steering committee called the **Biotechnology Programme Committee (BPC)**. The BPC is responsible for all policy matters. The Committee is supported by an intermediary organization i.e, the **Biotechnology Unit (BTU)** which is hosted by the Institute of Public Enterprise, Hyderabad. The other set of institutions are those engaged in research and extension activities with the support of the Programme.

Project Formulation, Implementation and Monitoring

The Programme has been using *pre-project formulation workshops* (PPFW) as a method of refining the priorities already identified and launching the specific projects for funding. PPFWs give enough scope for different stakeholders including farmers/farmer representatives to deliberate specific issues at length. Farmers' representatives explain their problems, constraints and expectations, which in turn forms the basis for further deliberations. *Group discussions* play a key role in identifying critical issues and problems to be pursued in the form of specific projects. The Programme also evolved certain guidelines for monitoring and evaluation of the projects and Programme. These are based on the *principles of participatory monitoring system* wherein the endusers are also consulted and their viewpoints are considered for further fine-tuning of the projects.

Progress Made

The Programme so far has made significant contributions in terms of evolving methodologies for problem identification, prioritization of interventions, people's participation and project formulation and monitoring. It succeeded in achieving greater commitment from the scientists towards achieving the identified goals and also succeeded in establishing sustainable networks with researchers on one hand and farming community on the other. Problems of resource poor farmers received particular attention in the Programme. Following the IBU approach the Programme has supported projects so far more than 60 specific projects. The projects deal with a range of technologies starting from simple, well established ones such as vermiculture, biofertilizers, biopesticides, botanical pesticides, biocontrol agents and tissue culture to high-tech biotechnologies such as genetic engineering.

The Programme established a number of tissue culture laboratories to produce and popularize qualitatively superior planting material of neem, teak, custard apple, tamarind, amla, karaya and a few important medicinal plants. The biofertilizers production cum extension unit established at one of the identified villages produces biofertilizers such as *Rhizobium, Azospirillum* and phosphate solubilizing bacteria and distributes to the farmers. The programme is also engaged in a big way in popularizing vermiculture technology by way of bringing awareness and training the rural youth and women in vermicompost production and application.

Considerable progress was achieved in the area of botanical and biopesticides. Technology development and transfer for biocontrol agents such as *Bacillus thuringiensis,* Baculovirus, *Trichoderma* etc. was also taken up for castor and groundnut crops. Extensive surveys were conducted in farmers' fields for isolation and identification of antagonistic fungi to manage castor wilt. Collection and screening of geographical isolates of semilooper Baculovirus and identifying potential strains of *Bacillus thuringiensis* was undertaken. Cost effective mass multiplication technology for *Bacillus thuringiensis* was developed. Farmers of the target villages were also trained in IPM practices.

Propagation of medicinal plants for general health and income generation also received due attention. Farmers were trained in preparation of important herbal products for treating some of the common ailments. The Programme is also engaged in the development and popularization of post harvest technologies. Economically feasible post harvest technologies for increasing shelf life of tomatoes and custard apple using biochemical approaches were standardized.

The projects on genetic transformation technology address the problems of biotic and abiotic stresses in the priority crops viz., sorghum, pigeonpea, castor and groundnut and also aim at capacity building of the individuals and institutions working in this area of research.

Livestock health and production is another important area which has made substantial progress in the Programme. The projects in this area were focussed on developing diagnostic kits and vaccines for livestock diseases. Intensive field trials were carried out for feed improvement through utilization of agro industrial by-products. An innovative programme on integrated livestock development was implemented in a big way covering all the aspects of livestock health and production viz. artificial insemination, vaccination, fodder improvement, feed supply, ram exchange, market linkages etc. Educated unemployed youth from the selected villages have been trained and provided with kits to attend animal health requirements in the villages. Realising the importance of public awareness on the introduction of new technologies like biotechnologies, Programme is also engaged in a big way in organizing systematic campaigns in biotechnology for different stakeholders in a project mode.

Impact Made

Convergence of indigenous knowledge with modern biotechnology

Individual projects are designed in such a manner that local people take active role in project conceptualization, formulation and implementation along with researcher. As such, projects address only need based technology. Which enabled sharing of indigenous knowledge of local people with modern biotechnology.

Resource mobilization

The Programme helped in resource mobilization and maximization of their utilization at the organization level. The APNLBP provides project based funds to the organizations through which they have been able to acquire their requirements like analytical equipments, instruments etc. and engage critical manpower needed. These funds along with existing resources of the host institute resulted in synergistic effect and enhanced capacities to develop biotechnologies. In some cases, these funds triggered creation of new institutional structures to promote R & D in biotechnology. In few cases, they helped in mobilizing funds from other donor agencies.

Social relevance of technology

Significantly IBU approach involving multistakeholders enhanced concern for social relevance of biotechnologies among researchers. Although initially it is hard for the scientists to follow this approach, as it is painstaking and time taking the researchers eventually appreciated its value.

Local capacity building in biotechnology

A significant contribution of the Programme has been creating capacities to analyse, prioritise and develop biotechnologies among partners. Capacity building takes place at three different levels - societal, organizational and individual. At societal level, the network provides opportunity for people to analyse their problems, prioritise them and seek solutions through biotechnologies. At the organizational level, the ability to undertake research on biotechnologies is enhanced through access to additional resources, information and knowledge. At the individual level, capacities are created through training, access to additional resources, information and interactions among network partners.

Dissemination of developed technologies

The results of the efforts of the Programme are disseminated to larger audience both within India and abroad through a Newseltter published quarterly in English and local language-Telugu. Independent of this newsletter, researchers also publish their results in peer-reviewed journals. Research results are also disseminated to farmers through periodic meetings and exposure visits. Many of the action oriented projects like tissue culture, biofertilizers, biopesticides, vermicompost, agroforestry etc. have already resulted in development of utilizable products. The local NGOs also played a crucial role in disseminating those technologies.

Quality and reliability of products

Both the technology developers and technology users are partners in the project right from inception and jointly undertake research studies and field demonstrations. This method has ensured delivery of qualitative and reliable products.

Demand driven biotechnologies

The combined efforts of participating partners in the projects resulted in demand driven biotech products such as planting material, obtained through tissue culture, biofertilizers, biopesticides, botanical pesticides, animal feeds and vaccines, vermicompost etc. Significant progress has been achieved in genetic transformation of plants, the higher end of biotechnology. Since the endusers are the partners in the project implementation, the spread of technology has been faster and effective as it moved from farmer to farmer.

Outlook for the future

In the coming years the Programme would focus on technology demonstration and transfer at field level and embark on new interventions in biotechnology. Gradually it is also proposed to integrate the activities of the programme with the other development programmes. The ultimate aim of the programme is to sustain the progress achieved and disseminate improved agricultural biotechnologies to help resource poor farmers.

4

Biotechnologies for Resource Poor Farmers of Zimbabwe

Julius T Mugwagwa
Biotechnology Trust of Zimbabwe, Harare, Zimbabwe

Introduction

Zimbabwe is a landlocked country, located in Southern Africa, just above the Tropic of Capricon and is 30 degrees east of the Greenwich Meridian. The Zimbabwean economy is largely based on agriculture, with agriculture (including forestry) contributing about 13% to 15% gross domestic product. Agriculture and forestry are important for exports contributing about 48% to forex earnings. An estimated 65% of the population derives their livelihood from agriculture, and agricultural activities produce about 60% of the raw materials. The agriculture sector is also an important and major market for industrial products such as fertilizers, agrochemicals and machinery.

The agricultural sector in Zimbabwe is dualistic, with a highly organized and developed commercial sector, having high levels of funding, mechanization and management systems. This sector consists of large tracts of farmland in high potential areas.

The Zimbabwe-Netherlands Biotechnology Programme

The Zimbabwean programme emerged in the early 1990's from the Special Programme on Biotechnology, funded by the Directorate General, International Cooperation of the Ministry of Foreign Affairs, the Netherlands. The programme was set-up with the objective of using biotechnology, in a need-driven, bottom-up and participatory manner, to address the agricultural production constraints faced by resource-poor farmers. A series of need identification and priority setting exercises at national and specific-target area levels came up with intervention areas for biotechnology. Two pilot districts were chosen, and these are Buhera and Hwedza, located in south-eastern central Zimbabwe which represents the bulk of the area where resource-poor farmers reside.

Table 4.1 Areas of Intervention for Biotechnology Under the Zimbabwean Biotechnology Programme

Area/Constraint	Intervention	Techniques Employed
Maize breeding	Improvement of drought-tolerance and insect resistance	Molecular-marker-assisted selection
Root and tuber crops	Provision of good quality planting material of sweetpotato through • Disease-elimination • Rapid propagation	• Tissue culture • DNA fingerprinting
Biological Nitrogen Fixation	• Integration of legumes in cropping systems	• Rhizobium inoculant technology
Inadequate on-farm animal feed	• Enhancing pasture legumes through inoculation • Silaging crop residues • Propagation of pasture legumes	• BNF • Fermentation technology • DNA fingerprinting
Poor animal health & reproduction rates	• Diagnosis development • Herd-health approach • Artificial insemination	• DNA- and antibody-based diagnostic kits • Artificial insemination
Shortage of relish and low income levels during dry seasons	• Mushroom production	• Tissue culture and other microbiological techniques • DNA fingerprinting
Capacity-Building	• MSc degree programme in biotechnology • Short-term fellowships • Farmers' training workshops	• All techniques in biotechnology
Other Potential Areas (specifically for drylands)		
Small ruminants production (disease-resistance and productivity selection)	• Breeding of goats and pigs	• DNA fingerprinting • Marker-assisted selection
Utilization of indigenous fruits	• Improving processing and post-harvest storage of indigenous fruits	• Fermentation technology
Addressing the problem of bird attack in sorghums	• Biochemical and genetic engineering of the flavonoid biosynthesis pathway	• Protein and DNA-marker techniques

Conclusion

Thus Zimbabwe agricultural biotechnology programme addresses the problems of resource poor farmers is two districts of Zimbabwe.

5

Molecular Marker Breeding for Drought Tolerance in Maize for Semi Arid Regions of Kenya

Kahiu Ngugi
Kenya Agricultural Research Institute, National Dryland Farming Research Centre, PO Box 340 Machakos, Kenya, E-Mail: eckngugi@net2000ke.com

Drought is one of the major causes for reduced crop production in the tropics. Globally, drought is the second most important cause of yield loss in maize, after insect pests (*Edmeades et al., 1992*). Grain yields in most crops are more susceptible to water deficits at flowering when the kernel is established. Maize yield is reduced to two to three times when water deficits coincide with flowering, compared with other growth stages (*Edmeades et al., 1992*). Grain yield of maize grown under severe stress at flowering and during grain filling is highly correlated with kernel number per plant (*Bolanos and Edmeades, 1996*). *Edmeades et al., 1999,* reported that decreasing the interval between anthesis and silking in maize increases yields in drought prone environments and this characteristic can be screened in large populations. In water limited environments, empirical breeding for improved yields has been slow due to the year to year variation in rainfall and within season variation in rainfall distribution in drylands. This has led to attempts to breed for specific phenological, morphological or biochemical characteristics that putatively improve yields in water limited environments (*Turner, 2001*). The use of molecular markers to select drought tolerance traits could speed up breeding of such complex trait as grain yield.

Plant breeders have long exploited the art and science of selection to improve crops for the various agronomic traits using the phenotype. In breeding crop cultivars, the most challenging task that breeders encounter is the difficulty of separating genetic effects imposed by the environment on the phenotype. Most traits of economic importance are quantitatively inherited and exhibit low heritabilities due to the large environmental effects. In these cases, it is difficult or impractical to determine at the phenotypic level whether or not specific gene or genes are present in a plant or cultivar. With the recent introduction of DNA molecular markers, it is now possible to select directly for the genotype using markers that co-segregate with the gene of interest. Molecular markers are characters whose inheritance can be followed at the DNA levels. They can be used indirectly to obtain information about the genetics of the traits of interest. Molecular markers can be expressed as DNA regions (genes) or DNA segments that have no known coding function. Examples of these markers include, Restriction Fragment Length Polymorphism (RFLP) and Polymerase Chain Reaction (PCR) markers such as Random Amplified Polymorphic DNAs (RAPDs), Sequence Tagged Sites (STSs), Simple Sequence Repeats (SSRs) and Amplified Fragment Length Polymorphisms (AFLPs).

Molecular markers have had their great use in linkage mapping of Quantitative Trait Loci (QTL). Effective QTL mapping depends on the association of phenotypic expression with markers. In major crops QTLs have been identified for yield components and for morphological and physiological characters associated with stress tolerance (*Ribaut and Poland; 2000*). Since most QTL for grain yield typically account for a small percentage of the phenotypic variance and exhibit high level of QTL X E interaction, it is generally deemed necessary to identify QTL for secondary traits which are correlated with yield under stress and whose variance increases under drought stress. In a marker assisted selection experiment using QTL for polygenic traits, *Ribaut et al., (1999)*, attempted to transfer five genomic target regions from a drought tolerant donor, Ac7643, to CML247, an elite but drought susceptible inbred line. After two back-crosses and two self-pollinations, the best genotype was fixed for the five target regions covering 12% of the genome. This paper briefly examines the advances so far made on the use of molecular markers in Kenya to select for drought tolerance in maize by identifying the pertinent QTLs.

Applications of Molecular Markers

RFLP markers are based on variation of genome DNA sequence. Unique sequences of DNA are cloned from the nuclear genome, and then used as radioactive or non-radioactive probes to detect homologous sequences in plant DNA which have been cut by various restriction endonucleases, separated on agarose gels and blotted on nylon filters. Alleles are identified by differences in the size of the restriction fragments to which the probes hybridize. Eventhough RFLP markers have been widely used, detection of RFLP by Southern blot hybridization may be laborious and time consuming, which may make this assay undesirable for plant breeding projects with high sample throughput requirements. Several types of PCR-based DNA markers have been developed including Random Amplified Polymorphic DNA,

RAPDs (*William et al., 1990; Welsh and McClelland; 1990*), sequence tagged sites, STSs, (*Olson et al., 1989; Talbert et al., 1994*), simple sequence repeats (SSRs) or microsatellites (*Beckmann and Soller, 1990; Roder et al., 1995*), and amplified fragment length polymorphisms, AFLPs (*Zabeau, 1993*).

Field Trials and Mapping Populations

A wide range of germplasm consisting of lines and their hybrids was screened under irrigated intermediate drought stress and severe drought stress conditions at the Kenya Agricultural Research Institute (KARI) substation Kiboko, in order to identify suitable parental lines. In the case of irrigated conditions, overhead sprinkler irrigation was applied throughout the crop growth cycle from germination to physiological maturity. For intermediate stress condition, the experimental plots were watered fully to sustain normal growth. Thereafter, one irrigation per week was applied for the next two weeks. Watering was completely withheld 14 days before flowering. It was expected that the realized grain yield under these conditions was about 80% of that under irrigated conditions. In the severe stress condition normal irrigation was applied only for the first 30 days. One more irrigation was applied 14 days later after which no further watering was done. About less than 40% of the potential grain yield was expected under these conditions. Twenty-five lines from lowland, tropical and subtropical germplasm obtained from International Maize and Wheat Improvement Centre (CIMMYT) were sown under irrigated, intermediate stress and severe stress conditions in an alpha-lattice (0.1) design with three replications. Forty hybrids arising mostly from these lines and including five Kenyan hybrids were also sown under irrigated, intermediate stress and severe stress conditions at the same time in alpha-lattice (0.1) design of three replications. In all experiments, plots were sown with three seeds per hill which were later thinned to one plant per hill. Each plot consisted of a 2.5 m row of 12 plants spaced at 20 cm between plants and 0.75 m between rows.

Mapping populations were generated from a cross between a CIMMYT drought tolerant line H16 and a drought susceptible high yielding Zimbabwe line K64R at Tlaltizapan, Mexico. Seven F1 ears were selfed in to produce F2 populations. From a labeled F2 population of 2000 plants sown in August 1998, 400 randomly selected ears were selfed to produce F3 families. Each of the 400 plants were crossed to two plants of testers, CML 202 and CML 311, 250 test-crosses were evaluated under the three watering regimes namely, irrigated, intermediate stress and severe stress levels at Kiboko substation of KARI.

Field Measurements and Data Analysis

In all the trials male flowering (MFLW) and female flowering (FFLW) were measured on individual plants. Anthesis to silking (ASI) was calculated as the difference between FFLW and MFLW family means. Grain yield (GRWT) and other yield components such as plant height (PLHT), ears per plant (EPP) per plot basis, were as follows :

- ♦ GRWT as the total weight of hand harvested, shelled grain dried at 90°C and converted to g / ha.

- ♦ EPP is the total number of ears in a plot divided by the number of plants in the plot.

- ♦ PLHT is the average length in centimeters of ten plants. Measured from the lower plant base to base of the tassels.

Adjusted means and genetic variance were calculated per family for each trait using the PROC MIXED procedure in SAS (*SAS Institute. 1988*). Using adjusted means of the F3 families, simple Pearson's correlation coefficients were calculated between the traits measured. Phenotypic, genotypic as well as broad-sense heritabilities were calculated.

RFLP and PCR Analyses

For the 130 RFLP markers used, genomic DNA was isolated from the two parental lines K64R and HI6 and also from 190 F3 plants. The DNA samples were quantified and digested with one of the two restriction enzymes (*EcoRI* and *HindIII*) separated in agarose gels (0.7%) and transferred to nylon membranes (MSI Magnagraph, Fisher Scientific) by Southern blotting. Labeled DNA probes (digoxigenin-dUTP) were used to detect polymorphism with Antidigoxigenin-alkaline phosphatase -AMPPD chemiluminescent reaction. Details of these protocols are given in Hoisington et al (1998). About 250 DNA probes from the University of Missouri, Columbia (UMC) and the Brookhaven National Laboratory (BNL) were used to screen the parental lines. The best polymorphic probes, 130 for drought tolerance were chosen to construct the linkage maps. In the F3 populations segregation ratios at each marker locus were tested by a chi-square goodness of fit test for the expected Mendelian segregation ratio, 1:2:1 for co-dominant loci and 3:1 for dominant ones. Hybridization blots were read and genetic data captured and verified by two readers using HyperMapdata, a software programme developed at CIMMYT.

For the 60 PCR markers used, DNA was isolated in a similar manner as for the RFLP markers using *Hoisington et al* (1998) protocols. Firstly parental lines were screened with about 200 primers using PCR protocols of *Hoisington et al* (1998) followed by electrophoretic analysis of the amplified products. Subsequently useful primers were selected and screened on the segregating 311 F3 individuals. A Chi-square analysis was performed on each PCR marker locus. This had been done for the RFLP marker locus to test for deviations from the expected gene frequencies of 0.5.

Mapping and QTL Estimation

Linkage maps were constructed using 311 F3 plants and 190 marker loci for population H16X K64R. Map positions of polymorphic loci were established by multipoint analysis using the computer programme "MAPMAKER" (version 3.0; log of odds (LOD) threshold = 3.0 and Theta threshold = 4.0) (Lander et al. 1987). Mapping of QTLs was performed using adjusted F3 family means for the traits measured in the field. The localization of QTLs for each trait was estimated by using "MAPMAKER/QTL" (version 1.1) software (*Lander and Eotstein, 1989*). Recombination frequencies were transformed into genetic distances using the Haldane mapping function.

QTL analysis was performed for H16 X K64R population on the subset of 280 F3 individuals using their phenotypic and genotypic data. Composite Interval Mapping (CIM) mixed model method of *Jiang* and *Zeng* (1995), which performs joint analysis of multiple environments and multiple traits was employed to estimate QTL and their genetic effects. The presence of a putative QTL was considered significant when the LOD score was larger than 2.5. Putative QTL with a LOD threshold under 2.0 were not considered. All the four models used to locate QTL on the chromosomes were tested. In model I, only those QTLs that gave a LOD score of 2.5 were considered. Model I was used to identify putative QTL regions by taking advantage of its high power of resolution. In model II, markers were used as cofactors and LOD score of 2.5 employed. In order to perform joint CIM for a single trait across the two seasons, the respective cofactor sets were combined. Models III and IV used all selected markers as cofactors flanking the target interval but employed minimum map distances (also referred as window sizes) of 20 and 30 cM respectively. Models III and IV were used to resolve QTL linked in coupling and repulsion phases. A QTL was declared present whenever a significant likelihood ratio (LR) peak was detected in model I and the LR exceeded the threshold in model II. If the LR was significant only under model I and a peak could not be detected in model II, a QTL was also declared. If the LR identified by models I and II was not confirmed by models III and IV, the hypothesis of the presence of a putative QTL at this position was rejected.

Adjusted entry means for GRWT, EPP, ASI, FFLW and MFLW were analyzed separately per season and then in a joint CIM across both seasons in order to determine the significance of QTL x Environment interactions.

Phenotypic and genotypic correlations between yield components

Phenotypic correlations presented in Tables 5.1 and 5.2 show that FFLW and MFLW are strongly and positively correlated as would normally be expected. Likewise ASI and FFLW are also positively correlated. ASI was however not correlated to MFLW as has also been shown in other studies (*Bolanos and Edmeades, 1993*) confirming that drought causes delays in silk emergence but has no or little effect on MFLW. These correlations were more or less similar under both severe stress and intermediate stress levels as shown by the correlations in Tables 5.3 and 5.4. In similar studies (*Ribaut et al., 1996*) it has been demonstrated that these correlations would be expected to be stronger under severe stress levels than under intermediate stress levels. These series of trials therefore may not have experienced a high degree of drought selection pressure. Table 5.5 confirms that there is indeed a genetic correlation between FFLW and MFLW and also between GRWT and EPP. This would imply that a QTL selected for any of these traits could be useful in the selection of other related traits. As would be expected, GRWT is highly negatively correlated to ASI implying that a shorter ASI would result in higher grain yields under severe stress as has been shown in other studies (*Ribaut et al., 1996*). Only FFLW and MFLW show high broad-sense heritabilities under both severe stress and intermediate drought levels (Table 5.6).

QTL Analyses

In the F3 population of H16 x K64R, one major QTL for GRWT located on chromosome 1 at about 184 cM was detected in the joint analysis across the two environments (Tables 5.7 and 5.9) when test-crossed to CML 202 and CML 311 (Table 5.8). The contribution of additive genetic variance in both environments was from H16 as indicated by the data on additivity. Another major QTL for GRWT was detected on chromosome 2 at 124 cM in CML202 test-crosses. A third QTL was detected on chromosome 4 at 180 cM in CML 311 test-crosses and had high significant LCD scores in models I and II. The two QTLs for GRWT in CML202 test-crosses contributed about 4.0 % genetic variance. These two QTLs had high significant LOD scores in models I, II and III. In model III, in the joint analysis across environments for both CML 202 and CML 311 test-crosses, the QTL X E interaction in chromosomes 1 and 2 were non-significant indicating that these QTLs are stable across environments (Tables 5.7 and 5.8). A minor QTL for GRWT was observed on chromosome 10 in the CML 311 test-crosses (Table 5.9).

In the joint analysis of CML 202 test-crosses (Table 5.7) MFLW showed high LOD scores when tested under all the models with a major QTL located on chromosome 9 at about 120 cM. However the QTL X E interaction was highly significant. In the CML 311 test-crosses, the QTL for MFLW was found on Chromosome 8 at 102 cM and gave non-significant QTL X E interaction. FFLW exhibited a QTL on chromosome 8 at about 100 cM with minimal QTL X E interaction. The MFLW QTL accounted for 28- 40% phenotypic variance whereas the FFLW QTL gave only 6.2% phenotypic variance. ASI gave one major stable QTL on chromosome 2 at 46 cM in the CML 202 test-crosses (Table 5.10) but two minor QTL chromosomes 3 and 10 at 162 cM and 80 cM respectively (Table 5.8). There were two major QTL for EPP on chromosomes 1 and 2 at 99 cM and 97 cM respectively in the CML 202 test-crosses (Table 5.7) and a minor one in the CML 311 test-crosses on chromosome 1 at 315 cM.

Table 5.1 Phenotypic Correlations of Yield Components in F3 Families Test-crossed to CML 202 under Severe Stress Conditions

Pearson Correlation Coefficients/ Prob/ R/ under H_o : Rho = 0 / N = 280

Trait	Phenotypic Correlation
1. FFLW vs MFLW	0.87 (0.0001)
2. ASI vs FFLW	0.65 (0.0001)
3. GRWT vs MFLW	− 0.35 (0.0001)
4. GRWT vs FFLW	− 0.33 (0.0001)
5. GRWT vs EPP	0.43 (0.0001)

Table 5.2 Phenotypic Correlations of Yield Components in F3 Families Test-crossed to CML 311 under Severe Stress Conditions

Pearson Correlation Coefficients/ Prob/ R/ under H_o : Rho = 0 / N = 280

Trait	Phenotypic Correlation
1. FFLW vs MFLW	0.88 (0.0001)
2. ASI vs FFLW	0.58 (0.0001)
3. GRWT vs MFLW	0.01 (NS)
4. GRWT vs FFLW	− 0.05 (NS)
5. GRWT vs EPP	0.51 (0.0001)

Table 5.3 Phenotypic Correlations of Yield Components in F3 Families Test-crossed to CML 202 under Intermediate Stress Conditions

Pearson Correlation Coefficients/ Prob/ R/ under H_o : Rho = 0 / N = 280

Trait	Phenotypic Correlation
1. FFLW vs MFLW	0.86 (0.0001)
2. ASI vs FFLW	0.44 (0.0001)
3. GRWT vs FFLW	− 0.12 (0.0001)
4. GRWT vs MFLW	− 0.014 (0.0001)
5. GRWT vs EPP	0.56 (0.0001)

Table 5.4 Phenotypic Correlations of Yield Components in F3 Families Test-crossed to CML 311 under Intermediate Stress Conditions

Pearson Correlation Coefficients/ Prob/ R/ under H_o : Rho = 0 / N = 280

Trait	Phenotypic Correlation
1. FFLW vs MFLW	0.71 (0.0001)
2. ASI vs FFLW	0.56 (0.0001)
3. GRWT vs MFLW	– 0.14 (0.0160)
4. GRWT vs FFLW	– 0.2 (0.0005)
5. GRWT vs EPP	0.48 (0.0001)

Table 5.5 Phenotypic Correlations of Yield Components in F3 Families Test-crossed to CML 311 under Severe Stress Conditions

Trait	Genotypic Correlation
1. FFLW vs MFLW	0.99 (0.0001)
2. ASI vs FFLW	0.21 (0.0001)
3. PH vs FFLW	0.08 (0.1704)
4. GRWT vs MFLW	0.30 (0.0001)
5. GRWT vs FFLW	0.24 (0.0001)
6. GRWT vs ASI	– 0.68 (0.0001)
7. GRWT vs EPP	0.91 (0.0001)

Table 5.6 Heritabilities of various Yield Traits of the Test - crosses under Intermediate Drought Stress (IS) and Severe Drought Stress Conditions (SS)

Env.	MFLW	FFLW	ASI	PLTH	EPP	GRWT
IS-SS-202	0.7062	0.6186	0.2590	0.2428	0.2034	0.2098
IS-SS-311	0.5281	0.4404	0.1658	0.1789	0.0463	0.0872

Table 5.7 Genetic characteristics of QTL involved in the expression of various yield components under IS and SS, F3 test-crossed to CML202.

Trial	Trait	Chromosome	QTL Position (cM)	Marker	LOD Score	QTL x E	Additivity	Phenotypic Variance IS	Phenotypic Variance SS
IS-SS	MFLW	9	55	5	15.15	7.08	−0.37	2.17	0.66
CML 202		9	129	3	12.56	12.03	−0.38	0.10	2.45
							28.7	26.2	
IS-SS	ASI	2	46	4	12.95	1.47	0.19	3.84	0.43
CML 202		8	89	6	10.04	9.90	0.02	0.86	2.44
							8.40	9.46	
IS-SS	EPP	1	98	6	21.94	0.01	0.23	2.21	0.38
CML 202		2	97	8			−0.03	11.01	16.18
IS-SS	GRWT	1	184	14	11.11	0.00	0.01	1.76	1.87
CML 202		2	124	9.11	0.08			2.14	0.06
								4.01	1.90

Critical values LOD = 3.0 = 13.81; LOD 2.5 = 9.21

Critical values; QTL x E, χ^2 0.052 = 3.84

IS -Intermediate Stress Level, SS – Severe Stress Level

Table 5.8 Genetic characteristics of QTL involved in the expression of yield components under IS and SS. F3 test-crossed to CML311

Trial	Trait	Chromosome	QTL Position (cM)	Marker	LOD Score	QTL x E	Additivity IS	Additivity SS	Phenotypic Variance IS	Phenotypic Variance SS
IS-SS	MFLW	8	102	7	19.96	1.68	– 0.54	– 0.25	7.69	3.49
CML 311									36.18	
									40.92	
IS-SS	FFLW	2	161	18	12.13	12.10	– 0.19	0.53	0.05	0.29
CML 311		8	102	7	14.54	4.64	– 0.54	– 0.15	6.18	0.82
									6.19	1.18
IS-SS	ASI	3	162	19	12.08	0.07	– 0.19	0.21	0.01	0.56
CML 311		10	80	6	14.03	3.28	– 0.41	– 0.13	3.28	1.12
									3.36	1.68
IS-SS	GRWT	1	201	16	10.95	0.91	0.11	0.20	0.27	2.43
CML 311		10	122	8	10.05	7.35	0.19	– 0.01	3.85	0.01
									6.26	4.18
IS-SS	EPP	1	315	19	9.33	0.80	11	23	1.60	1.39
CML 311							– 0.20	– 0.38		

Critical values LOD = 3.0 = 13.81; LOD 2.5 = 9.21

Critical values; QTL x E, χ^2 0.052 = 3.84

Table 5.9 Genetic characteristics of QTL involved in the expression of grain yield tested under Model I = a and Model II = b

Trial	Chromosome	QTL Position (cM)	Marker	LOD score
IS-SS CML 202	1 (a)	184	15	11.11
	1 (b)	184	15	10.02
IS-SS CML 311	4 (a)	180	10	9.40
	10 (a)	122	8	11.36
	4 (b)	180	10	9.81
	10(b)	121	8	11.85

Critical value LOD = 3.0 = 13.81; LOD = 2.5 = 9.21

Table 5.10 Genetic characteristics of QTL involved in the expression of ASI F3 test-crossed to CML202 tested under Model = a, Model II = b and ModelII = c

Trial	Chromosome	QTL Position (cM)	Marker	LOD score	QTL x E
IS-SS CML 202	2 (a)	47	4	11.36	
	2 (b)	47	4	13.78	
	2 (c)	46	4	12.95	1.47

Critical value; LOD = 3.0 = 13.81; LOD = 2.5 = 9.21

Critical values: QTL x E. F2 c^2 0.052 = 3.84

Conclusion

In this study, two major QTLs for grain yield and ears per plant were detected on chromosomes 1 and 2 across the intermediate and severe stress environments at 190 cM and 124 cM respectively. A QTL for ASI was identified in the same location as that of grain yield and ears per plant. Another QTL for grain yield and ASI was found on chromosome 4. Minor QTL whose LOD scores were below the threshold were detected on chromosomes 1, 3, 5 and 10.

As reported in other studies (*Ribaut et al., 1999*) grain yield QTLs have been found on Chromosomes 1 and 2. Considering that ASI, GRWT and EPP also showed strong phenotypic and genotypic correlations between themselves, it is an indication that they have linked QTL. It is clear that there are several genomic regions that are involved in the expression of more than one trait. Given the low heritability of grain yield under water stress environments, one may consider using either reduced ASI or increased EPP in an attempt to select for grain yield under water limited environments.

Although the degree of drought stress applied was not as strong as to allow for finer analysis of the genetic variance under water limiting conditions, the QTL for grain yield on chromosomes 1 and 2 showed no significant QTL X E interaction indicating that these QTLs were stable across environments. Strongly correlated traits such as female flowering and male flowering showed strong QTL X E interactions across the two environments. The QTL for ASI on chromosome 2 however gave non-significant QTL X E interactions an indication that this QTL was stable in the two environments. It is likely that this QTL accounts for reduced ASI whose alleles are contributed by the drought tolerant parent H16.

The complexity of selecting for grain yield under these circumastances is enhanced by the low phenotypic variance that the three QTLs exhibited. Hence the best marker assisted selection strategy would be to combine selection of grain yield QTL with selection for QTL key traits such as ears per plant and anthesis to silking interval.

References

Beckmann J.S. and Soller, M. (1990). Towards a unified approach to genetic mapping of eukaryotes based on sequence tagged microsatellite sites. **Biotechnology** 8: 930-932.

Bolanos, J. and Edmeades, G.O. (1993). Eight cycles of selection of drought tolerance in lowland tropical maize. II. responses in reproductive behavior. **Field Crops Research 31**:255-268.

Bolanos, J. and Edmeades, G.O. (1996). The importance of the anthesis-silking interval in breeding for drought tolerance in tropical maize. **Field Crops Research** 48, 65-80.

Edmeades, G.O., Bolanos J. and Lafitte, H.R. (1992). Progress in breeding for drought tolerance in maize. IN: Wilkinson D (Ed) Proc 47th Annual Com and Sorghum Res. Conf. ASTA, Washington, pp 93-111.

Edmeades, G.O., Bolanos, J., Banzinger, M., Ribaut. J.M., White, J. W., Reynolds, M.P. and Lafitte, H.R. (1998). Improving crop yields under water deficits in the tropics IN: Chopra, V.L., Singh, R.B. and Varma, A. (Eds) Crop productivity and Sustainability-Shaping the Future.Proceedings of the Second International Crop Science Congress, Oxford and IBH, New Delhi, pp. 437-451.

Hoisington, D., Khairallah, M. and Gonzalez-de-Leon, D. (1998). Laboratory Protocols: CIMMYT Applied Molecular Genetics Laboratory. Third Edition, Mexico, D.F. : CIMMYT.

Jiang, C. and Zeng, Z-B. (1995). Multiple Trait analysis of Gene Mapping for Quantitative Trait Loci. **Genetics** 140: 1111-1127.

Lander, E.S. and Botstein, D. (1989). Mapping Mendelian factors underlying quantitative traits uisng RFLP linkage maps. **Genomics** 121: 185-199.

Lander, E.S., Green, P., Abrahamson J., Barlow, A., Daly, M.J., Lincoln, S.E. and Newburg, L. (1987). MAPMAKER: an interactive computer package for constructing primary genetic linkage maps of experimental and natural populations. **Genomics 1**:174-181.

Olson, M., L. Hood. C. Cantor. and D. Botstein. (1989). A common language for physical mapping of the human genome. **Science 245**: 1434-1435.

Ribaut, J.M., Hoisington, D.A., Deutsch J.A., Jiang C. and Gonzalez -De-Leon D. (1996). Identification of quantitative trait loci under drought conditions in tropical maize. I. Flowering parameters and the anthesis -silking interval. **Theor. Appl Genet 92**:905-914.

Ribaut, J.M., Edmeades, G.O., Betran, F.J., Jiang C. and Banzinger, M. (1999). Marker assisted selection for improving drought tolerance in tropical maize in: O'Toole, J. and Hardy, B.(Eds) Proceedings of the International Workshop on Genetic improvement for Water-Limited Environments. IRRI, Los Banos, pp 193-209.

Ribaut, J.M. and Poland, D. (Eds). (2000). Proceedings of the Workshop on Molecular Approaches for the Genetic Improvement of Cereals for Stable Production in Water-Limited Environments. CIMMYT, EL Batan, Mexico.

Roder, M.S., Plaschke, S.U., Konig, A., Bomer, Sorrells, M.E, Tanksley, S.D. and Ganal M.W. (1995). Abundance, variability and chromosomal location of microsatellites in wheat. **Mol. Gen Genet 246**: 227-333.

SAS Institute Incorporated (1988). SAS language guide for personal computers. Edition 6.03, Cary, North Carolina, USA.

Talbert, L.E., N.K. Blake, P. W. Chee, T.K. Blake, and G.M. Magyar. (1994). Evaluation of "sequence-tagged-site" PCR products as molecular markers in wheat. **Theor Appl Genet 87**: 789-794.

Turner, N.C. (2001). Optimizing Water Use. IN : Nosberger, J., Geiger, H.H.and Struik, P.C. (Eds) Crop Science: Progress and Prospects. (2001) CAB International, pp 119-135.

Welsh, J. and M. McClelland. (1990). Fingerprinting genomes using PCR with Arbitrary primers. **Nucleic Acids Res 18**: 7213-7218.

Williams, J.G. K., A. R. Kubelik, K.J. Livak, J.A. Rafalski, and S. V. Tingey. (1990). DNA polymorphisms amplified by arbitrary primers are useful as genetic markers. **Nucleic Acids Res 18**: 6531-6535.

Zabeau, M. (1993). Amplified fragment length polymorphism (AFLP). European Patent Application 92402629.7.

6

Molecular Markers in Aid of Crop Improvement : Progress and Prospects in India

T. Mohapatra
National Research Centre for Plant Biotechnology, IARI, New Delhi 110012, India.

The ultimate difference between individual plants lies in the nucleotide sequence of their DNA. Detection of such differences employing various molecular biology techniques has led to development of molecular markers. Molecular markers follow simple Mendelian pattern of inheritance. They are stable and not influenced by developmental or environmental factors. Use of molecular markers to assist plant breeding has given birth to the sub-specialization of molecular breeding. This consists of development of molecular markers, construction of high density linkage maps, mapping and tagging of genes, and marker assisted selection. Molecular markers are based on two basic techniques: 1. Southern blot hybridisation (Southern 1975) and 2. polymerase chain reaction (PCR, Mullis et al., 1986).

Molecular Marker Systems

A survey of literature reveals many marker systems. All these are the result of modifications of the two basic techniques. In view of their importance and the extent of use, five of the molecular marker systems are described here.

Restriction Fragment Length Polymorphism (RFLP)

This is the first molecular marker system that was conceived and developed by Botstein et al. (1980) and is based on the Southern hybridisation technique. It employs restriction endonucleases, which recognize four to six nucleotide long sequences-the target sites, in the DNA molecules to be cleaved. It also utilizes single/low copy number DNA sequences as probes. The size (length) of the DNA fragments homologous to the probe DNA depends upon the location of target sites. Base changes can alter the sequences of the target site, thereby abolishing sites or creating new sites for a particular enzyme. DNA rearrangements such as deletions and insertions of sequences can also change the location of target sites. RFLP profiles are highly reproducible and by employing them it is possible to determine allelic diversity and frequency in a population.

Many RFLP markers are obtained by using DNA probes from different sources. For instance, probes can be prepared form cDNA clones from libraries obtained from different plant parts like leaf, flower, root, stem and even from callus. Small genomic DNA fragments derived by restriction enzyme digestion can be used as probes. Since a large number of DNA probes are available, there are practically no constraints on the number of RFLP markers that can be generated. However, this is not true in all crop species. It depends on the genome structure of the species concerned. For instance, the abundance of RFLP is very high in maize and Brassicas but relatively low in tomato and rice.

Highly repetitive DNA sequences called microsatellites (< 6 nucleotide long sequences repeated up to 100 times at a locus) and minisatellites (< 100 nucleotide long sequence repeated up to 1000 times at a locus) have also been used as probes in Southern hybridisation to generate highly polymorphic DNA profiles. Use of minisatellites for identification of human individuals has given birth to the concept of multi-locus DNA profiling, appropriately named as DNA fingerprinting. It is easy to identify crop varieties employing multi-locus DNA probes.

Random Amplified Polymorphic DNA (RAPD)

This marker system is based on polymerase chain reaction. It employs a single decamer primer of arbitrary sequence which is annealed to the template DNA typically at 37 °C. The RAPD profile is obtained by resolving the amplified products on an agarose gel. The variation in RAPD profile in the form of presence or absence of band results from variation in primer binding sites. A large number of RAPD markers can be generated by the use of a variety of decamer primers. A major limitation of this marker system is non-reproducibility due to low primer annealing temperature. However, the utility of a desired RAPD marker can be enhanced by sequencing its termini and designing longer (e.g. 24 nt) primers for specific amplification through PCR. Such modification that imparts the required reproducibility has been named as sequence characterized amplified region (SCAR).

Sequence Tagged Sites (STS)

This marker class was first conceived and used in human genome mapping and has been recently included in genome mapping in plants. STS markers are generated by an unmodified PCR. The primers are designed based on the sequence of cDNA, random genomic DNA and ends of large genomic DNA fragment cloned in cosmid, lambda or artificial chromosomes. Sequences flanking the microsatellites can also be used to generate PCR primers for assaying the variation in the number of tandem repeats in a given repeat motif at homologous sites. The STS marker system has all the advantages of PCR technique. A limitation of STS is that considerable effort and time are needed to generate DNA sequence data for designing suitable primers. Besides, the fragments amplified by the primers may not show length variation between individuals. This problem however, is circumvented by restriction digestion of the PCR products which detects polymorphism in the target site of a restriction enzyme internal to the two primer binding sites. This modification has led to a new marker system called cleaved amplified polymorphic sequence (CAPS).

DNA Amplification Fingerprinting (DAF)

DAF is another PCR based marker system which employs shorter (< 10 nt) primers of arbitrary sequence. The primer annealing temperature is also lower (30 °C) as compared to RAPD. The amplified products are separated on polyacrylamide gels and detected by silver staining. This gives a better resolution essential for proper visualisation and scoring of large number of bands usually obtained in OAF. The number of polymorphisms scored per run is high. Since the length of primer is short and primer annealing temperature is low, DAF suffers from the problem of non- reproducibility of some of the amplified fragments.

Amplified Fragment Length Polymorphism (AFLP)

The AFLP markers are generated by selective amplification of DNA fragments obtained by restriction enzyme digestion. High molecular weight DNA is digested by two restriction enzymes, one hexacutter (e.g. *Eco* RI) and one tetra cutter (e.g. *Mse* I). Adaptor molecules are ligated to the ends of DNA fragments. Two primers possessing sequence complementarity to the adaptors as well as few (typically 3) extra random nucleotides at their 3' ends are used for selective amplification of fragments employing PCR. The amplified products are separated on sequencing gels or even on ordinary PAGE and visualized by silver staining. Alternatively, the primers are labeled either by radioisotope or fluorescent dye so that the AFLP profile can be obtained by autoradiography or by using image analysis. The highest number of amplified products (50-120) is obtained through AFLP among all the DNA profiling systems. This increases the probability of detecting polymorphism many folds. The technique is, at present, lengthier and costlier than other PCR based techniques. It requires good quality DNA for ensuring complete digestion by enzymes because partial digestion of DNA results in non-reproducible variation in DNA profiles.

Construction of Molecular Maps

Construction of genetic linkage maps using morphological markers could not be initiated in most crop plants due to lack of sufficient number of markets. Morphological marker based map making was largely confined to maize and tomato. Even in these species, map construction was highly laborious, took many years and required several mapping populations since all morphological markers could not be obtained in a single cross. These maps contained limited number of markers and, therefore, could not be used for efficient mapping of target genes. With the availability of a large number of molecular markers, saturation mapping of plant genomes has become a reality. Construction of molecular genome map involves five major steps: (1) identification of divergent parents, (2) generation of a suitable mapping population, (3) identification of polymorphic probe enzyme combination and informative primers, (4) analysis of marker segregation in the mapping population, and (5) establishment of linkage.

Molecular genome maps have been constructed in almost all important crop plants. The number of markers employed to construct these maps and the marker density varies greatly. Most of these maps are based on RFLP markers, although in few crop species such as rice, RAPD, STS, and AFLP have also been utilized. Among the crop plants, the rice genome map (Harushima et al. 1998) contains the maximum markers (2275) and is considered most saturated. Significantly, this map has been made using a single F_2 population. Majority of these markers are derived from cDNA and thus are expressed sequence tags. Availability of such a genome map is facilitating identification of tightly linked flanking markers for genes of agricultural importance.

In India, molecular mapping was initiated in mustard (*Brassica juncea*) using RFLP markers and F2 population developed from an inter-varietal cross (Sharma et al. 1994), which was subsequently extended with additional markers to understand the extent of duplication of DNA sequences (Mohapatra et al 2002). Following selfing of F2 plants and single seed descent for 10 generations, a permanent mapping population of recombinant inbred lines (RILs) were developed and used to incorporate RAPD markers into the *B. juncea* genome map (Sharma et al 2002). This is the only genome mapping effort in the country.

Molecular Mapping of Genes

Availability of molecular markers and saturated linkage maps has enabled mapping of genes for qualitative as well as for quantitative traits. The qualitative traits show simple Mendelian pattern of monogenic inheritance. The quantitative traits, on the other hand, are polygenically inherited. Methods employed for the detection and mapping of loci for these two categories of traits are different. Mapping of genes for monogenetic traits in crop plants involve three distinct approaches : 1. use of a complete linkage map, 2. use of near isogenic lines and 3. bulked segregant analysis of F_2 population. The first approach essentially requires availability of a complete genetic linkage map. For the purpose of mapping a gene, markers are selected from all the chromosomes such that the maximum distance between any two adjacent markers does not exceed 20 centimorgan (cM). A gene mapping population (F_2 or BC_1) is created by crossing two parental lines having contrasting characteristics. The individual plants of this population are phenotyped for the target trait. Simultaneously, genotyping of these plants is carried out using markers which distinguish the two parental lines, in addition to being evenly distributed in the genome. Co-segregation of any of these markers with trait phenotype in the mapping population reveals linkage. This facilitates immediate localisation of the gene on the linkage map, since marker positions are already known. Although laborious, it is the most directed method of gene mapping.

The other two approaches do not require a linkage map. For instance, in case of the use of near-isogenic lines (which are developed by repeated backcrossing of the F_1 with the recipient parent and therefore, differ for a single gene), DNA markers distinguishing the two lines (e.g., resistant vs. susceptible) are first identified. Co-segregation of these markers with the trait is analysed in a mapping population, often derived from the cross of the two isogenic lines. Development of near isogenic lines, however, takes many years. This method of gene mapping therefore, is best suited to crops, in which intensive backcross breeding has been already practised to generate isogenic lines.

Use of F_2 population in conjunction with a technique called bulked - segregant analysis, is the most rapid way of mapping genes. In this approach, DNA from ten homozygous resistant F_2 plants are bulked to constitute a resistant bulk. Similarly, a susceptible bulk is made out of DNA from ten homozygous susceptible F_2 plants. DNA markers which distinguish these two bulks are considered to have linkage with the target gene. Eventhough, molecular linkage map is not a pre-requisite for its success, use of markers with known map position imparts greater efficiency to gene mapping in this approach.

Majority of the traits of agronomic importance are polygenically inherited. Such traits are traditionally characterised by applying the principles of quantitative genetics which enables estimation of the approximate number of loci affecting a character, average gene action, degree of interaction among polygenes and genotype-environment interaction. However, the magnitude of effect, inheritance and gene action in respect of any specific locus which is now commonly known as quantitative trait locus (QTL) and its interaction with other loci and environment remain unanswered.

The first attempt for identification of an individual QTL was made by Sax (1923). He reported a cross between two strains of *Phaseolus vulgaris*, one of which had large, coloured seeds and the other had small, white ones. The seed colour variation was due to a single gene difference, while seed size showed continuous variation. In F_2, the plants homozygous for the allele giving coloured seed had an average seed weight greater than that of the plants homozygous for the white colour allele. The average weight of the seed from plants heterozygous for the coloured allele was just about mid-way between the two. It was, thus, evident that the seed size was nearly proportional to the number of coloured alleles present in the F_2 individual. This suggested presence of linkage between the seed colour locus with one evidence in support of linkage of a major gene affecting flower colour with a QTL for flowering time in pea. The principle outlined in these

41

experiments was further extended and applied for mapping QTL in *Drosophila* (Thoday 1961). With the advent of molecular markers and construction of high density linkage maps, it is now possible to locate most of the QTLs for a trait segregating in a cross and determine their individual as well as interaction effects on trait expression. Precise identification of QTL, however, requires permanent mapping populations such as double haploids or recombinant inbred lines which can be replicated over seasons and locations for estimating environmental component of the total phenotypic variance.

Many genes of interest have been located precisely on the genome maps in several crop plants following these approaches. Initially, RFLP markers were used extensively for mapping and tagging of genes. Subsequently, RAPD was preferred over RFLP markers as soon as it was developed because of its simplicity and rapidity. There are several instances, where tightly linked (< 5 cM) markers located on either side of the target gene, have been identified. Prof. S.D. Tanksley and his colleagues, working at the Cornell University, Ithaca, USA were the first to use complete RFLP map of tomato and interval mapping for dissecting complex quantitative traits such as fruit mass and total soluble solid content into Mendelian factors (Paterson et. al. 1988). Based on information from complete molecular genome map, QTLs have been mapped for a variety of traits in many different plant species. This includes complex traits such as grain yield, drought tolerance and quantitative disease resistance. It has become possible to know the number and genomic distribution and type of gene action of QTL for the target traits. Interestingly the number of QTLs found to have major effects on trait expression is often only few (< 5). This has generated optimism for a directed manipulation of complex quantitative traits for which no selection criteria are available so far.

In India, molecular markers have been used to map genes for different traits mainly in mustard, wheat and rice. In mustard, Mukherjee et al. (2001) mapped a locus designated as *Ac2*(t) effective against Indian isolate of the white rust pathogen. Sharma et al. (1999) identified three loci for oil content in mustard seed based on segregation of RAPD markers in a recombinant inbred population. Sharma et al (2002) recently mapped two major QTLs influencing oleic acid level in *B. juncea* using both single factor analysis of variance and interval mapping. These two loci were located in 10.6 cM and 14 cM marker intervals respectively, and together explained 32.2% of the trait variance. Upadhyay et al. (1996) studied segregation of seed coat colour in an F_2 population of *B. juncea* and reported duplicate dominant gene action giving a phenotypic ratio of 15:1. Two RFLP markers flanking one of the interacting loci were identified. In a recent report, the seed coat colour trait was tagged using a combined approach of Bulked Segregant Analysis (BSA) and AFLP (Negi et al. 2000). In rice, genes for resistance to gall midge (at the ICGEB, New Delhi) and blast (at the MS University, Boroda) have been mapped using RAPD markers, which were converted to SCARs. Besides, mapping of genes for drought tolerance is underway at the University of Agricultural Sciences, Bangalore.

In wheat, genes for leaf rust resistance and grain quality have been mapped using molecular markers. Naik et al. (2001) identified RAPD markers for the leaf rust resistance locus Lr28. Prasad et al. (1999) suggested quantitative nature of Grain Protein Content (GPC) with several genes distributed throughout the genome. Recently, Dholakia et al. (2001) have demonstrated environmentally stable as well as specific loci associated with GPC and have shown the utility of DNA markers in studying G x E interactions affecting GPC. Gupta et al (1999) reported chromosomes IAS to be associated with thousand kernel weight (TKW) using STMS markers. AmmiRaju et al. (2001) further dissected the seed size and shape into its components such as seed length, width and density and showed ten markers to be associated with seed length and seed width and three markers for factor form density. The same study also demonstrated the chromosome regions 1D, 5B, 6A, and 6B for seed length, 2D and 6B for seed width and 1A and 2D for form density.

Marker Aided Selection (MAS)

Selection of a genotype carrying desirable gene or gene combination via linked marker(s) is called marker-aided selection. Breeders practice marker aided selection when an important trait, that is difficult to assess, is tightly linked to another Mendelian trait, which can be easily scored. For example, a gene for purple coleoptile color in some traditional rice varieties is closely linked to a gene that confers resistance to brown plant hopper (BPH). In a segregating population like F_2, about 95% of the plants showing purple coleoptile are found resistant to BPH. In this case, coleoptile colour is a morphological marker which is used to assist selection for BPH resistance. Morphological markers are however, limited in number, are specific to particular genotypes and are dominant. Morphological markers may also show tissue and developmental stage specific expression, pleotropy and even at times, adverse effect on plant growth, vigour and viability. Due to these reasons, morphological markers could not be of much use in MAS. Molecular markers do not suffer from these limitations and thus offer advantages over the morphological markers.

Molecular marker aided selection involves scoring for the presence or absence of a desired plant phenotype indirectly based on DNA banding pattern of linked markers on a gel or on autoradiogram depending on the marker system. The rationale is that the banding pattern revealing parental origin of the bands in segregants at a given marker locus indicates presence or absence of a specific chromosomal segment which carries the desired allele. This increases the screening efficiency in breeding programmes in a number of ways such as:

- The segregants can be scored at the seedling stage for traits that are expressed late in plant development. This includes traits such as grain quality, male sterility and photoperiod sensitivity.

- It is possible to screen for traits that are extremely difficult, expensive or time consuming to score and measure such as tolerance to drought, salt, mineral deficiencies and toxicity, root morphology, resistance to nematodes or to specific races or biotypes of diseases or insects.

- Selection can be practiced for several traits simultaneously which is difficult or even impossible by conventional means.

- Heterozygotes are easily identified and distinguished from the either homozygotes without resorting to progeny testing. This saves time and effort.

Specific plant breeding programmes in which marker assisted selection has been already utilised are: 1. gene introgression and elimination of linkage drag, 2. gene pyramiding and 3. development of heterotic hybrids.

Gene Introgression

Wild relatives of crops constitute a rich source of genetic variation which can be efficiently utilised for improvement of both qualitative and quantitative traits. In tomato, for example, genes resistant to *Meloidogyne incognita* and tobacco mosaic virus have been introduced from the related species *Lycopersicum*. Similar potential exists in wild and exotic germplasm for most crop plants. Molecular markers can aid introgression and impart greater precision. With the use of isozyme markers it has been possible to introgress nematode resistance into the cultivated tomato from one of its wild relatives. In an interspecific cross of tomato, it has been shown that selection for the recurrent parent markers along with the introgressed gene can result in major savings in the number of generations required for completion of back crossing procedure and in the amount of field space required to test the mature individuals.

It is often observed that the desirable genes such as those for disease resistance remain linked with undesirable weedy characteristics of the alien species. During gene introgression by back crossing, the linked undesirable gene also gets transferred to the recipient parent. This has been referred to as linkage drag. Even after 20 backcrosses, one expects to find a sizeable piece upto 10 cM of the donor chromosome still linked to the gene being selected. In most plant genomes, this segment is enough to contain hundreds of genes. This can, thus, result in a new variety modified for characters other than those originally targeted.

Molecular linkage maps provide a method to reduce the problems with linkage drag by allowing selection of individuals containing recombinant chromosomes which break linkage drag. It has been estimated that the use of molecular markers can reduce linkage drag at least 10 fold in a fraction of the time needed by traditional breeding (Tanksley et al.1989). In tomato, molecular linkage map has been used to identify varieties carrying minimal linkage drag around several disease resistance genes transferred from wild species.

Gene Pyramiding

Pyramiding of genes has been suggested as an effective way to provide durable form of disease and insect resistance in crop plants. Because of difficulty involved and the number of years it requires, development of lines carrying genes for resistance against several races of a single pathogen / insect or against different biotic stresses has not been very successful. Construction of such lines is much more efficient with molecular markers. Recently, successful pyramiding of four genes, *Xa4, xa5, xa13* and *Xa21* conferring resistance to four different races of bacterial leaf blight pathogen has been achieved in rice (Huang et al. 1997) by using molecular markers, which was otherwise impossible through conventional methods. Lines carrying different combinations of four genes for bacterial leaf blight (BLB) resistance, *Xa4, xa5, xa13* and *Xa21*, could be identified which were then used by different workers to develop BLB resistant cultivars. For instance, at the Punjab Agricultural University, Ludhiana, Singh et al. (2001) carried out marker aided pyramiding of these genes and developed BLB resistant rice cultivar PR106. This demonstrated that markers could be efficiently utilised to combine recessive and dominant genes together to develop improved cultivars. Similar works are underway on pyramiding of BLB and blast resistance genes at the Central Rice Research Institute, Cuttack and of gall midge resistance genes at the Indira Gandhi Agricultural University, Raipur.

Construction of Heterotic Hybrids

Molecular marker-based genetic analysis can be utilized to identify specific genomic regions containing QTLs for yield exhibiting dominance or over dominance gene action in the F_2 of a cross between two inbred lines. This will enable analysis of an existing cross in terms of the number of QTLs involved and the magnitude of their effect on the measured trait. Subsequently, additional inbred lines can be tested against standard lines for the presence of additional dominant or overdominant QTLs. These could then be incorporated into the standard lines by marker-based introgression. In this way the existing inbreds can be improved first and then used to develop hybrids for realizing higher heterosis than in the existing ones.

In maize, Prof. C. Stuber and his colleagues, working at the North Carolina State University, USA, have been successful in the above breeding scheme using molecular markers. QTLs for yield were first identified in inbred lines Tx 303 and oh 43, which were then transferred through use of molecular markers to inbreds B 73 and Mo 17 respectively. By crossing the improved B 73, with the improved Mo 17, a new hybrid [B 73 (I) x Mo (I)] was constructed which out yielded the best check by 15%.

Future Prospects

Complete integration of MAS with conventional plant breeding programmes demands consideration of two important factors: (a) size of population and

(b) cost. Plant breeding experiment requires screening of large segregating populations routinely over generations. Genotyping of large number of samples manually is an extremely difficult task. MAS to be practicable, should be amenable to automation that would allow handling of large number of samples. Development and use of PCR based markers such as STS and SCAR will be the key to success of MAS in crop improvement. As the technology develops and gets modified to analyse large number of samples, the cost will automatically go down. The investment in gene tagging, and selection based on molecular markers should be weighed against the overall cost and time involved in traditional breeding program. Even though, the cost of MAS is higher at the present level of estimation, its integration with traditional plant breeding is desirable because of the possibilities it offers.

References

Ammiraju JSS, Dholakia BB, Santra DK, Singh H, Lagu MD, Tamhankar SA, Dhaliwal HS, Rao VSP, Gupta VS, and Ranjekar PK (2001). Identification of inter-simple sequence repeat markers associated with seed size in wheat. **Theor. Appl. Genet.** 102, 726-732.

Botstein, D., White. RL., Skolnick, M and Davis, R W (1980). Construction of a genetic linkage map in man using restriction fragment length polymorphisms. *Am J Hum Genet.* 32314-331.

Dholakia BB, Ammiraju JSS , Santra DK, Singh H, Katti MV, Lagu MD, Tamhankar SA, Rao VS, Gupta VS, Dhaliwal HS and Ranjekar PK .(2001) Molecular marker analysis of grain protein content in wheat (*T. aestivum*). **Biochem. Genet.** (In press)

Gupta PK, Balyan HS, Prasad M, Varshney RK, Roy JK, Harjit Singh, and Dhaliwal HS (1999) Molecular markers for some quality traits in wheat IN: National Symposium on Frontiers of Research in Plant Sciences. P 14, Dec 2 - 4, 1999, Calcutta, India.

Harushima, Y., Yano, M., Shomura, A., Sato, M., Shimano, L Kuboki, Y, Yamamoto, 1., Lin, S., Antonio, B.A., Parco, A., Kajiya, H., Huang. N., Yamamoto, K., Nagamura, Y., Kurata. N., Khush, G.S. and Sasaki, 1. (1998) A high density rice genetic linkage map with 2275 markers using a single F, population. *Genetics.* 148. 479-494.

Huang. N., Angeles, E.R. Dorningo, I., Magpantay, G, Singh, S. Zhang. G., Kumaravadivel, N., Bennett, *I.*, and Khush, G.S. (1997) Pyramiding of bacterial blight resistance genes in rice. marker assisted selection using RFLP and PCR. *Theor. Appl. Genet.* 95.313-320

Mohapatra T, Upadhyay A, Sharma A and Shanna RP (2002) Detection and mapping of duplicate loci in *Brassica juncea*. **J. Plant Biochem. Biotech.,** 11: 37-42

Mukherjee AK, Mohapatra T, Varshney A, Sharma R and Sharma RP (2001) Molecular mapping of a locus controlling resistance to *Albugo candida* in *Brassica juncea*. **Plant Breeding,** 120 (6): 483-487

Mullis, K.B" FaloDNA, F., Scharf, S., Saiki, R" Horn, G., and Erlich, H. (1986) Specific enzymatic amplification of DNA *in vitro'* the polymerase chain reaction. *Cold Spring Harbor Symp Quanti BioI.* 51.263-273.

Naik S, Gill KS, Prakasa Rao VS, Gupta VS, Tamhankar SA, Pujar S, Gill BS, Ranjekar PK (1998) Identification of a STS marker linked to the *Aegilops speltoides*-derived leaf rust resistance gene Lr28 in wheat. **Theor. Appl. Genet.,** 97: 535-540

Negi MS. Devic M, Delseny M and Lakshmikumaran (2000) Identification of AFLP fragments linked to seed coat colour in *Brassica juncea* and conversion to SCAR marker for rapid selection. **Theor. Appl. Genet** 101: 146-152

Paterson, A.H., Lander, E.S., Hewitt, *I.D.*, Peterson, S., Lincoln, S.E. and Tanksley, SD (1988) Resolution of quantitative traits into Mendelian factors by using a complete linkage map of restriction fragment length polymorphism. **Nature** 335: 721-726.

Prasad M, Varshney RK, Kumar A, Balyan HS, Sharma PC, Edvard KJ, Singh H, Dhaliwal HS, Roy JK, Gupta PK (1999) A microsetellite marker associated with a QTL for GPC on chromosome arm 2DL of bread wheat. **Theor. Appl. Genet** 99: 341-345

Rasmusson, I.M (1933). A contribution to the theory of quantitative character inheritance. *Hereditas.* 18, 245-261.

Sax, K. (1923). Association of size differences with seed coat pattern and pigmentation in *Phaseolus vulgaris, Genetics* 8: 552-560.

Sharma A, Mohopatra T and Sharma RP (1994) Molecular mapping and character tagging in *Brassica juncea*. Degree, nature and linkage relationship of RFLPs and their association with quantitative traits. **J. Plant Biochem Biotech.** 3: 85-89.

Sharma R, Mohapatra T, Mukherjee AK, Krishanpal and Sharma RP (1999) Molecular markers for seed oil content in Indian mustard. **J. Plant Biochem. Biotech.,** 8: 61-64

Sharma R, Aggarwal RAK, Kumar R, Mohapatra T and Sharma RP (2002) Construction of RAPD linkage map and localization of QTLs for oleic acid level using recombinant inbreds in mustard. **Genome,** 45 (3): 467-472

Singh S, Sidhu JS, Huang N, Vikal Y, Li Z, Brar DS, Dhaliwal HS and Khush GS (2001) Pyramiding three bacterial blight resistance genes (*xa5, xa*13 and *Xa*21) using marker assisted selection into indica rice cultivarPR106. **Theor. Appl. Genet.**, 102:1011-1015

Southern, EM. (1975) Detection if specific sequences of DNA fragments separated by gel electrophoresis. *J. Mol Biol* 98: 503-517.

Tanksley, S.D., Young, N.D., Paterson. A.H., Bonierbale, M.W. (1989). RFLP mapping in plant breeding. new tools for an old science. **Biotechnology**. 7: 257-264.

Thoday, I.M (1961). Location of polygenes. *Nature* 191: 368-370.

Upadhyay A, Mohapatra T, Pai RA and Sharma RP (1996). Molecular mapping and character tagging in Indian mustard (*Brassica juncea*) II. RFLP marker association with seed coat colour and quantitative traits. **J. Plant Biochem. Biotech.**, 5: 17-22.

7

Molecular Markers and Marker Assisted Selection in Dryland Crops

T. Gopalakrishna
Nuclear Agriculture and Biotechnology Division
Bhabha Atomic Research Center, Trombay, Mumbai 400 085

One of the main goals for plant breeders would be to develop genotypes that have relatively high productivity in water-stress environment. This aspect would be very relevant to India where 60% of the presently cultivated area is rainfed. This approximately translates into 97 million hectares and supports about 40% of the population. It would be interesting to note that nearly 90% of coarse cereals, 90% of total pulses, 80% of oil seeds and about 65% of cotton produced in the country come from rainfed areas. Globally, 69% of all cereal is rainfed, including 40% of rice, 66% of wheat, 82% of maize and 86% of coarse grains. Not all problems posed before a breeder can be solved using the conventional approach. The conventional selection and breeding procedures have evolved over several decades to an effective and satisfactory methodology. However, some traits are difficult to tackle by conventional approach since they are complex and often devoid of visible markers that makes selection process difficult. Few of such characters are tolerance to abiotic stresses, grain quality, heterosis for yield etc. The response of plants to drought is complex and diverse. It is unrealistic to expect that drought tolerance will be a monogenic trait. DNA based molecular markers can immensely help the plant breeder in such situations.

DNA Markers

DNA markers are alleles of loci, which show polymorphism due to nucleotide base sequence variation. DNA markers score over morphological and biochemical markers since the environment does not influence them and are not restricted to any particular tissue or stage in the life cycle of the plant. A number of marker systems have been developed. Some of the popular marker systems are random amplified polymorphic DNA (RFLP) (Welsh and McClelland 1990), restriction fragment length polymorphism (RFLP) (Beckmann and Soller 1986), microsatellites or simple sequence repeats (SSRs) (Morgante and Olivieri 1993), inter simple sequence repeats (ISSRs) (Zietkiewicz *et al.*, 1994), amplified fragment length polymorphism (AFLP) (Vos *et al.*, 1995) etc. Of these different techniques only RFLP is based on DNA-DNA hybridization while the other methods are based on the polymerase chain reaction (PCR) (Saiki *et al.*, 1988). PCR based techniques are less time consuming and require only minute quantities of DNA thus making it very useful. Based on the extent of polymorphism one observes that SSRs score over the other methodologies. RAPDs are easier to perform compared to RFLP, AFLP and SSRs. RAPD results have problems of reproducibility and informativeness. Thus each method has its own advantages and disadvantages. The choice of the system depends on the objective of the study and perhaps the facilities available in the laboratory.

DNA Marker studies at BARC

At BARC, RAPDs were used for estimating genetic purity in cotton hybrids NHH-44 and DCH-32 (Pendse *et al.*, 2001; Yadav *et al.*, 2001) and for varietal identification (Ranade and Gopalakrishna 2001; Bhagwat *et al.*, 2001). However, for tagging the disease resistance gene in groundnut AFLP is being used (Bhagwat, pers. comm.).

Markers in Plant Breeding

Genetic markers are not new to plant breeding but DNA markers are (Table 7.1) novel tools that are used in plant breeding. The application of DNA markers in plant breeding is numerous and only a few are mentioned here. They can be used for

- Confirming breeding line identity
- Establishing hybrid identity
- DNA fingerprinting and germplasm characterization
- Genetic diversity estimation in germplasm
- Construction of genome maps
- Comparative genomics
- Tagging of agronomically useful genes
- Map based cloning and
- Marker assisted selection

Table 7.1 Evolution of genetic markers

Type	Period	Year
Morphology and cytology	Early Genetics	1900 to 1950s
Isozyme and proteins	Pre rDNA	1960s to mid 1970s
RFLP and minisatellites	Pre PCR	mid 1970s to 1985
RAPD, microsatellites, ESTs, STSs and AFLP	PCR age	1986 to 1995
DNA sequence and proteomics	Cyber Genetics age	1996 to present

At Bhabha Atomic Research Centre (BARC) DNA based molecular markers were used for genetic purity determination in cotton hybrids as well as for DNA fingerprinting of varieties. Genome mapping is a procedure in which the markers are put in order indicating the distances between them in terms of recombination values. So, markers in a single chromosome would belong to one linkage group. These linkage maps can be used for studying interrelationships as well as in marker assisted selection (MAS). Linkage maps are available for some of the dryland crops. They include pearl millet (Liu et al., 1994), sorghum (Xu et al., 1994), mungbean (Menacio-Hautea et al., 1993), cowpea (Menedez et al., 1997), chickpea (Simon and Muehlbauer, 1997) and cotton (Shappley et al., 1996). One of the applications of linkage maps is in comparison with distantly related taxa where the information from one is used to predict the linkage in the other taxa. Synteny or gene order can help in predicting the location of genes of interest in related species. Although genes have not been isolated on the basis of synteny, it holds a lot of promise for map-based cloning.

The construction of genetic linkage maps has been useful in tagging genes with molecular markers (Mohan et al., 1997). However, many agronomically useful traits such as yield, resistance to biotic and abiotic stresses, quality and maturity are controlled by a large number of loci. Such loci are termed as quantitative trait loci (QTLs). QTLs are different from the polygenic traits that show a continuous phenotypic variation.

Plant breeders attain their objective by using strategies that can be described as having three steps namely,

1. Identification of genetic variations among individuals that show different characteristics.

2. Making genetic crosses and obtaining recombinants between genotypes that exhibited different characteristics in other words develop genotypes with new (or improved) sets of attributes and

3. Accurate selection of these not so common recombinants from a large population.

Marker Assisted Selection (MAS)

MAS is a breeding strategy where in an indirect selection procedure is applied. Instead of selecting for the gene itself the marker that is closely linked to the gene controlling the trait of interest is used for monitoring. MAS can significantly accelerate breeding efforts by providing new alternatives to difficult-to-achieve goals by classical breeding techniques alone. After establishing that a marker lies close to a gene of interest, the breeder can assay for the presence of the marker in the segregating population and determine with confidence whether the gene of interest is present or absent at a very early stage of growth, even before the gene had a chance to express the trait in the phenotype. Since selections can be made accurately at seedling stage the number of individuals that must be grown to maturity can be significantly reduced. Such a selection would also be more accurate since it is based on the genotype than solely depending on phenotype, which could be due to either genotype or its interaction with the environment. An advantage of MAS is not only to select individuals with the help of markers linked to the trait of interest but also discard individuals for backcross breeding for indirect selection of desirable gene(s) and minimizing the linkage drag from the donor parents.

MAS will also play an important role in the development of transgenic plants. In the asexual transformation method of transferring genes, only certain genotypes that are amenable to tissue culture methods are used. These genotypes might not be the most popular ones with the cultivators. So after developing the transgenic plant, backcross breeding has to be resorted to, to transfer the transgene to a popular variety. MAS could come into a major role here. In the case of polygenic traits MAS is more efficient than the traditional selection especially when the trait under study has low heritability. In the early generations MAS gives greater genetic gain. However, in later generations the gains by traditional selection may be equal or even surpass that of MAS. MAS enables breeders to improve the efficiency of breeding when an important trait that is difficult to screen is tightly linked to a marker that can be easily determined.

The essential requirements for MAS in a plant breeding programme are that (i) markers (preferably codominant) should be tightly linked to the trait of interest (ii) a means to screen rapidly a large population should be available and preferably based on DNA amplification and (iii) the screening should be highly reliable and economical.

MAS for biotic stresses

In breeding for resistance against biotic stresses, a segregating population is screened either at hot spots or under artificially created stress condition. These procedures although have yielded results are time consuming and expensive. Further, there are always the 'escapes'. Availability of markers for resistance genes will help in identifying plants perhaps even at seedling stage without subjecting to pathogen or pest attack. So even breeding can be done in the absence of the pathogen or the pest. Many genes controlling resistance to pathogen or pest have been tagged with molecular markers (Table 7.2). But MAS would be useful for tracing QTLs and major genes where screening procedure is labour intensive or difficult. MAS would also be very useful in gene pyramiding (Kelly et al., 1995). The usefulness of MAS specially for breeding traits controlled by QTLs has been reviewed (Dudley 1993) and is beyond the scope of this article.

Table 7.2 Selected list of crops and their resistance genes that have been tagged[1]

Crop	Disease	Resistance gene
Sorghum	Downy mildew	
	Head smut	Shs
Chick pea	Wilt	
Mungbean	Bruchids	
	Powdery mildew	
Groundnut	Root rot nematode	Mae, Mag

[1] For details of references please see Mohan et al., (1997) and Kumar (1999)

MAS for abiotic stresses

Application of molecular marker technologies has helped to understand the underlying physiological parameters that control a plant's response to abiotic stress. This understanding has been of immense importance in programmes on plant improvement. This is because the plant phenotype is the result of the expression of many different and diverse physiological as well as biochemical processes. Most physiological assay systems cannot be applied for a screening procedure in a breeding program, as the methodologies are too complex. In plant breeding the dissection of a complex trait or physiological parameter into simpler components is of importance. Application of genomic technologies for drought tolerance has listed hundreds of genes as likely candidates based on gene function (Seki et al., 2001). Some of these genes are listed in table 7.3. It is essential to identify amongst these genes those that are involved in drought tolerance per se. It should be possible to develop a consensus plant with the necessary candidate genes imparting drought tolerance which is a major constraint in the dryland crops.

Table 7.3 Selected drought inducible genes identified by cDNA microarray

Gene	Accession	Coded protein no. of cDNA	Drought (2h) (ratio[a])
rd29A	D13044	Hydrophilic	6.4 + 2.8
erd10	D17714	Group II LEA protein	6.0 + 2.3
rd17	Ab004872	Group II LEA protein	6.4 + 1.6
rd19A	D13042	Thiol Protease	2.8 + 0.3
rd22	D01113	Unidentified seed protein	2.2 + 0.2
erd4	AB039928	Membrane protein	2.6 + 0.8
FL5 - 1A9	AB050542	Nodulin like	2.9 + 1.2
FL5 - 94	AB050551	Enolase	2.0 + 1.5
FL5 - 3J4	AB050562	Hsp-DNA J homolog	2.8 + 0.3
FL5 - 27	AB044405	Cyst PI homolog	2.2 + 0.4
FL5 - 3E18	AB050568	Aquaporin homolog	2.0 + 0.3
FL5 - 3A15	AB050589	Ferritin	2.8 + 0.9
FL3 - 2C6	AB050571	Thioredoxin	2.3 + 0.5
FL2 - 5G7	AB050551	Catalase 3	2.4 + 0.7
FL5 - 90	AB050575	β amylase	nd

[a] Florescence intensity (FI) of cDNA for dehydration condition
FI for unstressed condition

Recent work on response to stresses suggests that manipulating gene regulation and signal transduction may increase the plant adaptation to abiotic stresses. Identification of genes involved in stress-induced transcription factors, involved in the initiation phase of the stress response is one such example. The discovery of elements like dehydration responsive element (DRE) and ABA responsive element (ABRE) in *Arabidopsis* and the identification of their transcription factors are recent developments. A multidisciplinary research effort involving breeders, physiologists and molecular biologists is needed for efficient germplasm.

MAS at what price ?

Like any other research activity the use of MAS in plant breeding programmes will be governed by the cost factor. RFLP techniques are very expensive when the consumables as well as the necessary trained technical staff are considered. RAPD markers on the other hand will suit the budget of most laboratories. But PCR with arbitrary primers is associated with the problem of low reliability. Increasing the reliability of PCR reaction and simple DNA extraction procedures to suit large-scale screening have to be addressed first. At BARC the DNA extraction method that works well with legume seeds was developed (Krishna and Jawali, 1997). DNA markers that can be scored as present or absent without the electrophoresis step would reduce the cost of analysis. The development of sequence characterized amplified region (SCAR) would be a prerequisite for this. The costs of chemiluminescent and radioactivity based RFLP and RAPD have been compared (Ragot and Hoisington 1993). However, the extrapolation of this calculation to MAS will not be appropriate.

Current Status and Future Perspective

Although linked molecular markers have been identified for agronomically important traits in many crops, their use for MAS in actual plant improvement or breeding programmes is not yet visible (Gupta and Roy, 2002). The functional genomics and its allied technologies offer a new breeding tool. The current challenge is how to integrate these into the current breeding programs to improve MAS strategies for developing crops. In India most of the molecular marker work has been and is being done by non-plant breeders. We need to have a synergistic approach of involving of new technologies into plant breeding programs.

References

Beckmann JS and Soller M (1986) Restriction fragment length polymorphisms in plant genetic improvement. **Oxford Survey of Plant Molec Cell Biol.** 3, 196-250.

Bhagwat AS, Anantharaman R, Pereira SM and Gopalakrishna T (2001) Identification of polymorphism by random amplified polymorphic DNA (RAPD) in groundnut (*Arachis hypogaea* L.). **Plant Var and Seeds** 14, 119-124.

Dudley JW (1993) Molecular markers in crop improvement : Manipulation of genes affecting quantitative traits. **Crop Sci.** 33, 660-668.

Gupta PK and Roy JK (2002) Molecular markers in crop improvement: Present status and future needs in India. **Plant Cell, Tissue and Organ Culture** 70, 229-334.

Kelly JD, Afanador L and Haley SD (1995) Pyramiding genes for resistance to bean common mosaic virus. **Euphytica** 82, 207-212.

Krishna TG and Jawali N (1997) DNA isolation from single half seeds suitable for random amplified polymorphic DNA analyses. **Anal Biochem** 250, 125-127.

Kumar LS (1999) DNA markers in plant improvement: An overview. **Biotech Adv** 17, 13-182.

Liu CJ, Witcombe JR, Pittaway TS, Nash M, Hash CT, Busso CS and Gale MD (1994) An RFLP-based genetic map of pearl millet (*Pennisetum glaucum*). **Theor Appl Genet**. 89, 481-487.

Menacio-Hautea D, Kumar L, Danesh D and Young ND (1993) A genome map for mungbean (Vigna radaiata (L.) Wilczek) based on DNA genetic markers (2n = 2x = 22). In : O' Brien JS, editor. Genetic Maps: A compilation of linkage and restriction maps of genetically studied organisms. Cold Spring Harbor, NY: Cold Spring Harbor Laboratory Press, pp. 6259-6261.

Menedez CM, Hall AE and Gepts P (1997) A genetic linkage map of cowpea (*Vigna unguiculata*) developed from a cross between two inbred, domesticated lines. **Theor Appl Genet**. 95, 1210-1217.

Mohan M, Nair S, Bhagwat A, Krishna TG, Yano M, Bhatia CR and Sasaki T (1997) Genome mapping, molecular markers and marker assisted selection in crop plants. **Molec Breed** 3, 87-103.

Morgante M and Olivieri (1993) PCR-amplified microsatellites as markers in plant genetics. **Plant J** 3, 175-182.

Pendse R, Malhotra S, Pawar SE and Gopalakrishna T (2001) Use of DNA markers for identifying inbreds and hybrid seeds in cotton (*Gossypium hirsutum L.*) **Seed Sci and Technol**. 29, 503-508.

Ragot M and Hoisington DA (1993) Molecular markers for plant breeding: comparison of RFLP and RAPD genotyping costs. **Theor Appl Genet** 86, 975-984.

Ranade R and Gopalakrishna T (2001) Characterisation of blackgram (*Vigna mungo* (L.) Hepper) varieties using RAPD. **Plant Var and Seeds** 14, 227-233.

Saiki RK, Gelfand DH, Stoffel S, Scharf R, Higuchi R, Horn GT, Mullis KB and Erlich HA (1988) Primer-directed enzymatic amplification of DNA with a thermostable DNA Polymerase. **Science** 239, 487-491.

Seki M, Narusaka M, Abe H, Kasuga M, Shinozaki KY, Carninici P, Hayashizaki Y and Shinozaki K (2001) Monitoring the expression pattern of 1300 *Arabidopsis* genes under drought and cold stresses by using full length cDNA microarray. **The Plant Cell** 13, 61-72.

Shappley ZW, Jenkins JN, Watson CE Jr, Kahler AH and Meredith WR (1996) Establishment of molecular markers and linkage groups in two F_2 populations of upland cotton. **Theor Appl Genet**. 92, 915-919.

Simon CJ and Muehlbauer FJ (1997) Construction of a chickpea linkage map and its comparison with maps of pea and lentil. **J Hered**. 88, 115-119.

Vos P, Hogers R, Bleeker M, Reijans M, van de Lee T, Hornes M, Frijters A, Pot J, Peleman J, Kupier M and Zabeau M (1995) AFLP: a new technique for DNA fingerprinting. **Nucl Acids Res**. 23, 4407-4414.

Welsh J and McClelland M (1990) Fingerprinting genomes using PCR with arbitrary primers. **Nucl Acids Res.** 18, 7213-7218.

Xu GW, Magill CW, Schertz KF and Hart GE (1994) A RFLP linkage map of Sorghum bicolor (L.) Moench. **Theor Appl Genet** 89, 139-145.

Yadav M, Ranade R, Vaidya UJ and Gopalakrishna T (2001) Genetic - purity determination of cotton hybrid DCH-32 by random amplified polymorphic DNA (RAPD). **Plant Var and Seeds** 14, 35-40.

Zietkiewicz E, Rafalski A and Labuda D (1994) Genomic fingerprinting by simple sequence repeat (SSR)-anchored Polymerase chain reaction amplification. **Genomics** 20, 176-183.

8

Use of Marker Assisted Selections (MAS) in Crop Breeding Programmes

P. N. Rajesh and Vidya Gupta
Plant Molecular Biology Unit, National Chemical Laboratory, Pune 411 008, Maharashtra.

Plant breeding has been one of the most successful technologies developed in modern agriculture. Many of the limitations of important plant breeding methods have been because of lack of technical infrastructure for conducting genetic analyses. Marker assisted technology is one of the potential tools to significantly strengthen and complement various breeding programs. Usage of genetic markers, which are heritable entities that are associated with economically important traits by breeders for phenotypic selection, can produce dramatic changes in breeding populations (*Staub et al. 1996*). This technology will not be able to replace the time tested methods but can definitely enhance their efficiency.

Various Markers Useful in MAS

Although DNA sequencing is a straight forward approach for identifying variations at a locus, it is very expensive and laborious (*Joshi et al. 1999*). DNA markers are being used as versatile tools for investigating different aspects of plant genomes. Various types of molecular markers are available and they can be classified into the following types.

- ♦ **Single or Low Copy**

    (a) Restriction fragment length polymorphism (RFLP)

    (b) RFLP converted into PCR based

    - Sequence Tagged Sites (STS)

    - Allele Specific Associated Primer (ASAP)

    - Expressed Sequence Tag (EST)

    - Single Strand Conformation Polymorphism (SSCP)

◆ **Repetitive DNA**

(a) Microsatellite and Minisatellite

(b) PCR based repetitive DNA markers

    - Sequence Tagged Microsatellite Sites (STMS)

    - Directed Amplification of Minisatellite-Region DNA (DAMD)

    - Inter Simple Sequence Repeat (ISSR)

◆ **Arbitrary Sequence markers**

(a) Random Amplified Polymorphic DNA

(b) Modification in RAPD

    - DNA Amplification Fingerprinting (DAF)

    - Arbitrary primed PCR (AP-PCR)

    - Sequence Characterized Amplified Region (SCAR)

    - Cleaved Amplified Polymorphic Sequences (CAPS)

    - Random Amplified Microsatellite Polymorphism (RAMPO)

    - Amplified Fragment Length polymorphism (AFLP)

State of Art of the Technology

RFLP markers are co-dominant markers, which are capable of detecting coupling phase of DNA molecules as DNA fragments from all homologous chromosomes are detected. They are reliable and informative markers in linkage analysis and breeding as they can easily determine if a linked trait is present in a homozygous or heterozygous state in an individual. Also, information highly desirable for recessive traits. The limitations of this system are requirement of large amount of DNA, usage of radioactive isotope or expensive non-radioactive kit and also their inability to detect single base changes. These constraints have been overcome by converting them into PCR based markers. When such types of markers are linked to some specific traits, they can be easily integrated into plant breeding programs for marker assisted selection of the trait of interest.

The repetitive DNA regions play an important role in absorbing mutations in a genome and inherited mutations are vital in evolution and also these sequences constitute 30%-90% of the genomes of all species. Thus repetitive DNA and mutational forces together form the basis for a number of marker systems that are useful for several applications in plant genome analysis. Also, these sequences have been detected in the expressed transcripts also, repetitive DNA based markers are capable of detecting polymorphism at RNA level also.

RAPD markers are single primers of arbitrary nucleotide sequences which detect nucleotide sequence polymorphisms. This marker system gained its significance because it can detect variation in any part of the genome. Although it is simple to perform as it is PCR based, the constraints in this technique are its dominant nature and reproducibility. Hence there is a necessity of modification in this system, which resulted in developing DAF, SCAR, CAPS, RAMPO markers (*Joshi et al. 1999*).

Achievements made so far (Crops and Traits)

Molecular markers have made greater impact in phylogenetic relationships, cultivar identification, genetic diversity, parentage determination, and marker assisted selection. These markers have been widely used in tagging many agronomically important traits and in developing linkage maps in various crops. In rice, markers have been utilized to identify a number of QTLs for cooking and eating quality (*Tan et al. 1999*), in soybean for seed oil and protein content (*Lee et al. 1996*), in maize for grain yield, protein and starch concentration (*Stuber et al. 1987, Goldman et al. 1993*) and in wheat for kernel traits, kernel hardness and flour viscosity (*Campbell et al. 1999, Sourdille et al. 1996*). Our lab, in specific, has contributed in this area of research immensely and the identified markers have the potential for utilization in marker assisted selection. In chickpea, we have identified markers for *Fusarium* wilt, *Ascochyta* blight resistance and also for other agronomically important traits such as double podding (*Ratnaparkhe et al 1998; Santra et al. 2000; Rajesh et al. 2002*). In addition, markers have been used in analyzing heterosis in Indian elite chickpea cultivars and to determine genetic relationship among various cultivars and wild species (*Sant et al. 1999*). In rice and pearl millet, it has been used to evaluate phylogenetic relationship and hybrid performance and heterosis prediction (Chowdari et al. 1998). Marker has been identified for male sterility (TGMS) in rice from our lab. In wheat, we have identified markers for quality traits such as bread making quality, grain protein content, grain size and grain hardness, other traits like yellow berry tolerance, leaf and stem rust disease resistance (*Ammiraju et al. 2001; Dholakia et al. 2001*). It has been used for diversity analysis in tetraploid wheat too. Also we have performed diversity studies among natural populations collected from various regions of our country using molecular markers. In papaya, sex-determining factor has been tagged and also the marker has been patented (*Parasinis et al. 2000*). Markers have been used for diversity analysis of various pathogens such as *Ascochyta rabiei, Fusarium oxysporum, Xanthomonous oryzae* (*Rajebhosale et al, 1997*), *Neovossia indica* (Karnal bunt) (*Datta et al, 2000*).

Prospects and Constraints

Molecular markers have been looked upon as tools for a large number of applications ranging from localization of a gene to improvement of plant varieties by marker assisted selection. They have also become extremely popular for phylogenetic analysis adding new dimensions to the evolutionary theories. Genome analysis based on molecular markers has generated lot of information and a number of databases are now available for its preservation and popularization. The markers can also be utilized in genome fingerprinting, genome mapping, population genetics, taxonomy, plant breeding and diagnostics. Marker assisted breeding programs have been estimated to reduce the time-to-market by 50%-70% (*Tanksley et al. 1989; Schneider 1997*). The use of markers to study genotype X environment interactions and genetically dissect complex traits has gained enormous importance recently. In order to apply molecular breeding to large-scale breeding programs, automation technologies must be introduced. Hence, the constraint is mainly the cost of Marker Assisted Selection (MAS), which can be high when compared to classical phenotypic selection. Hence the cost-benefit relationship has been more critically evaluated in differing marker systems (*Staub et al, 1996; Lee 1995*). Also the lack of response to MAS was due to antagonistic effects of genome regions responsible for the yield and quality traits observed.

In summary, molecular markers and associated technologies can assist in map construction, the analysis of the molecular and genetic basis of quantitative and qualitative traits and most importantly, in disease resistance breeding. Since laboratory costs associated with MAS applications are decreasing recently and more effective and efficient molecular markers are being developed, MAS might have potential for selection of characters such as yield components in agronomic and horticultural crops.

References

Ammiraju JSS, Dholakia BB, Santra DK, Singh H, Lagu MD, Tamhanker SA, Dhaliwal HS, Rao VS, Gupta VS, Ranjekar PK (2001) Identification of ISSR markers associated with seed size in wheat. **Theor. Appl. Genet.** 102; 726-732

Campbell KG, Bergman CJ, Gaulberto DG, Anderson JA, Giroux MJ, Hareland G, Gulcher RG, Sorrells ME, Finney PL (1999) Quantitative trait associated with kernel traits in a soft X hard wheat cross. **Crop Sci** 39: 1184-1195

Chowdari KV, Venkatachalam S, Davierwala AP, Govilla OP, Gupta VS, Ranjekar PK (1998) Hybrid performance and genetic distance as revealed by (GATA) 4 microsatellite and RAPD markers in pearl millet. **Theor. Appl. Genet.** 97: 163-169.

Datta R, Rajebhosale MD, Dhaliwal HS, Singh H, Ranjekar PK, Gupta VS (2000) Intraspecific genetic variability analysis of *Neovossia indica* causing Kamal bunt of wheat using repetitive elements. **Theor. Appl. Genet.** 100: 569-575

Dholakia BB, Ammiraju JSS, Santra DK, Singh H, Katti MV, Lagu MD, Tamhanker SA, Rao VS, Gupta VS, Dhaliwal HS, Ranjekar PK (2001) Molecular marker analysis of protein content using PCR based DNA markers in wheat. **Biochemical genetics** 39: 325-338

Goldman IL. Rocheford TR, Dudley JW (1993) Quantitative trait loci influencing protein and starch concentration in the Illinois term selection maize strains.

Joshi SP, Ranjekar PK, Gupta VS (1999) Molecular markers in plant genome analysis. **Current Science** 77; 230-240

Lee M (1995) DNA markers and plant breeding programs. **Advances in agronomy** 55: 265-343

Lee SJ, Bailey MA, Mian MAR, Carter Jr. TE, Shipe ER, Ashley DA, Parrot W A, Hussey RS, Boerma HR (1996) RFLP loci associated with soybean seed protein and oil content across populations and locations **Thoer. Appl. Genet** 93: 649-657

Parasnis AS, Gupta VS, Tamhankar S, and Ranjekar PK (2000). A highly reliable sex diagnostic PCR assay for mass screening of papaya seedlings. **Molecular Breeding** 6: 337-344.

Rajebhosale MD, Chowdhari VK, Ramakrishna W, Tamhanker S, Gnanamanickam SS, Gupta VS. Ranjekar PK (1997) DNA fingerprinting of Indian isolates of *Xanthomonas oryzae* pv. oryzae. **Theor. Appl. Genet.** 95: 103-111

Rajesh PN, Tullu A, Gil J, Gupta VS, Ranjekar PK and Muehlbauer FJ (2002). Identification of an STMS marker for the double podding gene in Chickpea. **Theor. Appl. Genet.** (in press).

Ratnaparkhe MB, Santra DK, Tullu A, Muehlbauer FJ (1998) Inheritance of Inter simple sequence repeat polymorphism and linkage with Fusarium wilt resistance gene in chickpea. **Theor. Appl. Genet.** 96: 348-353

Sant VJ, Patankar AG, Gupta VS. Sarode ND, Mhase LB, Sainani MN, Deshmukh RB, Ranjekar PK (1999) Potential of DNA markers in detecting divergence and in analyzing heterosis in Indian elite chickpea cultivars. **Theor. Appl. Genet.** 98: 1217-1225

Santra DK, Tekeoglu M, Ratnaparkhe MB, Gupta VS, Ranjekar PK, Fred J. Muehlbauer. (2000) Identification and Mapping of QTLs conferring Resistance to Ascochyta blight in Chickpea. **Crop Sci.** 40:1606-1612.

Schneider (1997) Marker-assisted selection to improve drought resistance in common bean. **Crop Sci.** 37: 51-60

Sourdille P, Perretarrt MR, Charmet G, Leroy P, Gantier MF, Joudier P, Nelson JC, Sorrells ME, Bernard M (1996) Linkage between RFLP markers and genes affecting kernel hardness in wheat. **Theor. Appl. Genet.** 93: 580-586

Staub JE, Serquen FC, Gupta M (1996) Genetic markers, map construction and their application in plant breeding. **Hort. Science** Vol. 31 (5); 730-740

Stuber CW, Edwards MD, Wendel JF (1987) Molecular marker facilitated investigation of quantitative trait loci in maize. II Factors influencing yield and its components traits. **Crop Sci** 27: 639-648

Tan YF, Li JX, Yu SB, Xing YZ, Xu CG, Zhang Q (1999) The three important traits for cooking and eating quality of rice grains are controlled by a single locus in an elite rice hybrid, Shanyou 63. **Theor. Appl. Genet** 99; 642-648

Tanksley SD, Young ND, Paterson AH, Bonierbale MW (1989) RFLP mapping in plant breeding: New tools for an old science. Biotechnology 7: 257-264. **Theor. Appl. Genet.** 87:217-224

9

Molecular Markers in Rice Breeding

N.P. Sarma* and R. M. Sundaram
Directorate of Rice Research*, Hyderabad 500 030
e-mail : npsarma@lycos.com

Selection for desired trait(s) has been the hallmark of all plant breeding activities since the beginning of improvement of plants in agriculture. Plant breeders have used phenotype as the basis along with statistical methods to select superior segregants. The recent developments in molecular genetics and the advent of molecular markers has broadened the canvas of plant breeding, as they can be used as new selection tools to add precision to the art and science of plant-breeding. The opportunity to select desirable lines based on genotype rather than phenotype is really an exciting proposition to a plant breeder. Even though marker assisted selection (MAS) is expected to a play prominent role in crop improvement, particularly rice breeding, examples of successful and practical application are not many baring the cases of pyramiding the genes for biotic stress resistance in rice. The discovery that differences in endonuclease restriction sites could be used as markers to construct genetic maps in the late 1980s brought about a paradigm shift in rice breeding and genetics, when researchers began to substitute morphological and isozyme markers with DNA based molecular markers. The first Restriction Fragment Length Polymorphism (RFLP) map of rice was developed in 1988 at Cornell University (McCouch et al. 1988) and soon high-density linkage maps based on RFLPs were developed by others (Causse et al. 1994, Kurata et al. 1994). This quickly led to the vision of using the simple, abundant and inexpensive DNA markers in rice breeding. Polymorphism in the nucleotide sequence is the basis for these DNA markers to serve as selection tools. Such polymorphisms are revealed by techniques such as Restriction fragment length polymorphisms (RFLPs), Random amplified polymorphic DNA (RAPD), Amplified fragment length polymorphisms (AFLPs), Microsatellite or simple sequence repeat (SSR) polymorphism, etc. DNA markers like any other genetic markers are linked (i.e., located near) to the locus where the gene(s) controlling the trait of interest reside and co-segregate along with the trait of interest across generations. Genetic maps have been constructed in many crop plants using these markers and rice geneticists too have used these markers to map genes controlling quantitative and qualitatively inherited traits.

The utility of DNA markers is dependent to a large extent by the techniques that are used to reveal the underlying DNA polymorphism. They are generally classified as DNA-hybridization based markers and Polymerase chain reaction (PCR) based markers. In the first category are RFLPs which are visualized by hybridization of restriction enzyme digested DNA to a labelled probe of known sequence or origin. PCR based markers involve *in-vitro* amplification of specific DNA sequence or locus by using either specifically designed or arbitrarily chosen random primers. A plethora of markers are now available for choicest use. As each of these markers have different advantages and inherent limitations, the question of what markers to use depends on the specific purpose. For example trait introgression, qualitative and quantitative trait mapping, germplasm characterization or diagnostics have different technical challenges and demand specific requirements. The suitability of a marker is determined by several considerations such as ease of assay, ability to discriminate between individuals, the frequency of occurrence of the marker (abundance) and more importantly the type of marker – Co-dominant or Dominant. Though there is nothing like a universal ideal marker, SSR markers are generally considered as markers of choice in rice as they offer several advantages – abundance, high polymorphism, relatively uniform genome coverage, ease of detection and co-dominant inheritance. Further, for routine application of DNA marker technology to a large breeding population, automation in DNA isolation, polymerase chain reaction and electrophoresis steps are much desired.

Current status of application of molecular markers

Since the first RFLP linkage map of rice was constructed, several major genes and QTLs have been mapped. A review of the major genes tagged and mapped in rice to date reveal that more than half of them confer biotic stress resistance, understandably due to their economic importance. MacKill and Ni (2000) have listed all the major genes tagged and mapped in rice. A list of the traits along with total number of genes so far tagged and mapped as per latest reports is given in Table 9.1. More and more genes are being added year after year due to the concerted efforts of rice geneticists. It is also further revealing that majority of the genes conferring resistance to biotic stresses are assigned to Chromosomes 11 and 12, relative to the size in Centimorgans. Incidentally, Chromosome 11 and 12 share duplicate segments, a fact which has been established by molecular analysis (Kurata et al 1997).

Table 9.1 Rice genes so far tagged and mapped (2003)

Trait	No. of genes tagged	No. of genes mapped
Blast resistance	15	15
Bacterial leaf blight resistance	9	8
RiceTungro Virus resistance	1	1
Rice Yellow MosaicVirus resistance	1	1
Gall midge resistance	8	8
Brown Plant Hopper resistance	3	2
White Backed Plt. Hopper resistance	2	1
Fertility restoration	4	4
Thermosensitve Genetic Male sterility (TGMS)	4	3
Wild Abortive Cytoplasmic Male Sterility (WA-cms)	3	3

In molecular marker mapping, PCR based markers – AFLPs and SSRs are increasingly used. The mapping populations used for this purpose are either derived from doubled haploid populations of cross – Azucena x IR64, Recombinant Inbred Lines (RILs) of the cross - Nipponbare x Kasalath, Interspecific crosses - *O. sativa x O. longistaminata* and many more such crosses. The type of population (whether F_2, F_3, RILs, DH lines etc.) to be used in mapping studies depends on the trait attempted for tagging. For example, for qualitatively inherited traits, populations at early segregating generations (for eg., F_2, F_3) are equally good as advanced generation lines (for eg., RILs, DH lines), whereas for QTL-marker linkage analysis RILs and DH lines are preferred over early generation segregants.

Application opportunities in rice genetics and breeding

The important applications of molecular markers in rice breeding and genetics are:

(i)   Determining the allelism of gene(s) conferring identical phenotypes

(ii)   Use in Marker Assisted Selection

(iii) Map based Cloning of genes

(iv)   Tracking the introgression of transgene

Marker Assisted Selection (MAS)

There has been enough excitement about marker-assisted selection in Plant Breeding, since the time first DNA markers were developed. Selection based on genotype rather than phenotype, ascertaining plant genotype at juvenile stage or even seeds, simultaneous screening for multiple genes/traits in a single plant, accelerated recovery of recurrent parent genome in backcross breeding programmes are just a few attractions of MAS for the plant breeders. Prior to the advent of DNA marker technology, the idea of uncovering the loci controlling complex multigenic traits was thought to be impossible. Adoption of MAS by breeders will depend on cost, convenience and reliability when applied on the scale typical of plant breeding efforts.

Though MAS has become prominent in the field of plant breeding, a general rule is that if the gene of interest can be scored easily, it is uneconomical to resort to MAS.

Markers offer real advantage when

(i)   the trait phenotype is difficult or expensive to score

(ii)   when the trait is controlled by several genes conferring similar phenotypes

(iii) in breeding programmes involving recessive genes

(iv)   to select against donor genome while introgressing desirable genes from wild species or agronomically inferior genotypes.

MAS is not effective if the marker(s) is not tightly linked to the gene of interest. This largely depends on the marker system, mapping population used as well as on the quality of phenotyping. Phenotyping of quantitative traits can be more accurate with large sized populations, multiple replications in diverse environments, appropriate genetic and statistical analyses, and independent verification/validation through parallel and alternate populations at different environments. Other limitations, which affect the utility of MAS in plant breeding are lack of polymorphism between the parental lines (particularly in crosses between closely related parents usually resorted to in breeding programmes), cost and time consuming DNA extractions, PCR, gel electrophoresis and analytical procedures.

Plant breeding considerations of Marker assisted selection

Success in integration of DNA marker technology in plant breeding depends on several considerations such as

(i)   Choice of marker

(ii)   Ease of marker assay

(iii) Extent of polymorphism

(iv)   Dependability (Repeatability)

The choice of marker system depends on the objective, skill and facilities available with plant breeder. Markers such as RFLPs, AFLPs and RAPDs require high quality DNA. Furthermore, compared to PCR based techniques, RFLPs demand larger quantity DNA per sample and involve laborious, time consuming and costly assaying procedures. AFLPs though combine the advantages of RFLP and PCR requires higher skills, cost of assay and usully involves extensive handling of hazardous radioisotopes. For direct application by plant breeder, SSRs (microsatellites) are simple to use particularly in rice, since more than 5000 of these have been already developed and mapped (McCouch et. al. 2002). For other crops also, SSR markers are being identified and mapped by many public and private sector initiatives (http://gramene.org). SSRs are relatively simple to analyze, can be automated, are of high informative value, mostly monolocus and lowest in terms of cost per genotype and primer and hence are considered as best for use in rice breeding.

One major problem that limits the effectivity of MAS is false positives. If the marker and the target gene are not tightly linked, crossing over may cause segregation between them resulting in false positives. Inaccuracy in gene mapping is also the other reason for false positives. To avoid such pitfalls, fine mapping or high-resolution genetic maps or identifying tightly linked and flanking markers using larger population sizes are advocated. Markers developed based on the sequences of the gene(s) as exemplified in the case of Xa21 are of course the best.

Another important consideration is the cost of marker technology itself. As things stand, it is relatively expensive for the development of molecular markers and application of MAS in small laboratories is limited due to the high cost of material and supplies involved as well as demanding expertise and lab infrastructure required. It is anticipated that along with the growth of indigenous biotech supply industry, the cost of materials and reagents will ultimately come down. Of course, there are examples for which MAS application benefits are likely to outweigh the expenses. If MAS facilitates to accomplish something that is not possible through conventional methods, cost involved is more than justified. An excellent analysis of cost-benefit ratio of MAS has been illustrated by Dreher et al (2000) using the case study of introgression of quality protein in maize through traditional method and MAS.

As plant breeding is a game of numbers, DNA marker technology should cope with thousands of segregants per cross and many such crosses per season and should offer ways to meet fully this demand. Rice breeders generally practice pedigree breeding and/or as the situation demands, backcross breeding (eg., disease resistance, transfer of cms and restorer genes). Pedigree breeding involves multiple parents as donor sources for different traits. The segregating material generated needs to be screened simultaneously by breeders, pathologists and entomologists. Each F_2 segregant is a unique recombinant in itself and due to non-replicative nature of the material; breeders are confronted with the crucial problem of sharing the seeds at the same generation. Markers if available for such traits, are the best means to circumvent the problem by sharing of DNA from individual segregants and screening for multiple gene recombinants (MGRs) simultaneously.

Current applications in rice breeding

DNA markers are being used in rice breeding

(i) for traits difficult to phenotype (measure or screen)

(ii) for more durable resistance through gene pyramiding.

(iii) in accelerated recovery of recurrent genome in backcross breeding.

(iv) in transfer of one or more recessive genes for stress resistance/tolerance.

(v) targeted introgression of QTLs from wild relatives and land races into elite cultivars

(vi) tracking he introgression of transgene(s) in segregating populations.

The traits that can be measured only after flowering are natural candidates of choice for MAS, for eg. grain quality traits (like amylose content, presence of Beta-carotene in the grain), presence of thermosensitive genetic male sterility gene (TGMS gene) can be identified at seedling stage using markers (Ayers et al. 1997, Lang et al. 1998). Similarly, if the trait is observed only in the test cross progeny of individual plants (eg., trait controlled by a recessive gene), the best option is to employ MAS. Two good examples are recessive bacterial leaf blight resistance genes xa5 and xa13 and restorer gene for cms. The ability to identify the traits in early growth phase confers the advantage of pre-flowering selection for the choice of the line in further backcrossing without the loss of generation time.

Molecular breeding for durable resistance

Pyramiding different genes conferring resistance to diseases and pests is a practical approach for ensuring broad-spectrum and durable resistance. Different resistance genes confer resistance to different races, isolates, biotypes of the pest/pathogen as the case may be. When several genes confer the same phenotype, markers have been used to select for the gene pyramids.

Two good examples of MAS are

(i)   To combine three blast resistance genes in an elite line (Hittalmani et al 1996)

(ii) Pyramiding of Bacterial Leaf Blight (BLB) genes, *Xa4, xa5, xa13* & Xa21 (Huang et. al 1999, Singh et. al 2001).

The latter example illustrates the advantage of marker-assisted selection of recessive genes (*xa5* & *xa13*) in backcross programme without resorting to cycle of selfing thereby saving time. Three cases of successful deployment of DNA markers in rice improvement carried out at the Directorate of Rice Research, Hyderabad are described below.

DNA markers in Hybrid purity testing

Considering the importance of seed purity estimation and its maintenance in hybrid rice technology, a simple, rapid and reliable technology has been developed by DRR in collaboration with the Centre for Cellular and Molecular Biology (CCMB), Hyderabad for assessment of purity of seeds of hybrid rice and also the parental lines (Yashitola et al. 2002). The technique involves the use of simple, easily assayable PCR based markers like SSRs and Sequence tagged sites (STSs) markers for purity estimation. The methodology involves germination of a representative seed sample (~400 seeds) representing a seed lot for about 3-5 days, separation of the growing portions of individual seedlings from endosperm contaminants, DNA extraction using a quick protocol from each seedling, use of extracted DNA in a PCR assay involving SSR and/or STS markers, gel electrophoresis of the amplified fragments, gel documentation and analysis of the banding pattern to identify the impurities among the seedlings assayed. The entire assay right from DNA isolation to gel analysis takes just about 4-5 hours and can be carried out even in a moderately equipped laboratory. Thus this DNA marker based assay can serve as an alternative to the conventional assay through Grow-Out Test (GOT) commonly used in seed industry for seed purity estimation. An elaborate database of marker amplification pattern in different cms, restorer and other lines has been developed at DRR for use in such purity tests (Naveen Kumar, Unpublished). Another assay for distinguishing the cytoplasmic male sterile (cms) lines from maintainer and restorer lines has also been developed by CCMB-DRR involving the use of cms-mitochondria specific STS primers (Yashitola et al. 2004). Together, these assays are helping hybrid rice breeders and seed companies to rapidly and reliably monitor parental and hybrid seed purity.

Tagging, mapping and utilization of genes for gall midge resistance in rice

Rice gall midge is one of the most serious insect pests of rice in India. No reliable chemical control methodology is known for the pest and deployment of resistant varieties is considered the only feasible option for its management. More than 8 gall midge resistance genes (*Gm1, Gm2, gm3, Gm4, Gm5, Gm6* etc.) have been identified in rice. Even though many resistant varieties with single gene conferred resistance are being grown widely, the emergence of new biotypes of the insect pest, resulting in breakdown of resistance has complicated pest management strategies. Pyramiding of more than one resistance gene in the background of an elite cultivar is considered as the best strategy for durable resistance. DRR has tagged and mapped a dominant gall midge resistance gene, *Gm1* from the cultivar W1263 with the help of microsatellite and RAPD markers. After identification of a tightly linked RAPD marker OPB2$_{(476)}$, chromosome specific SSR markers were used to fine map the gene locus on chromosome 9. The SSR markers on chromosome 9S -RM219, RM316, RM444 which have been identified to flank the gene locus are being used for introgression of the gene from the donor Kavya into the genetic background of popular rice cultivar Samba Mahsuri. In a similar such effort the earlier identified markers for the gall midge resistance gene *Gm2* (Nair et al 1995) have been validated in different genetic backgrounds and tightly linked flanking markers viz. RM317 and F8 have been identified for targeted introgression of the gene into elite cultivars.

Pyramiding of bacterial leaf blight resistance genes into elite rice cultivars

Bacterial leaf blight (BLB) is a devastating disease of rice caused by a bacterial pathogen, *Xanthomonas oryzae pv oryzae*. Like gall midge, no chemical control methodology is available to control the disease and deployment of resistant varieties is the only option for its management. For development of durable BLB resistance in the background of elite rice cultivar Samba Mahsuri, a collaborative marker-assisted backcross breeding programme was initiated involving DRR and CCMB. One dominant (*Xa21*) and two recessive (*xa5 & xa13*) BLB resistance genes were introgressed into the genetic background of Samba Mahsuri with the help of tightly linked PCR based DNA markers. Locus specific SSR markers spread across the entire rice genome were used for accelerated recovery of the recurrent parent genome at each backcross generation, ultimately reducing the backcrossing to only BC_4. Analysis of the selfed progeny at BC_4F_2 and BC_4F_3 clearly showed that the selected plants are not only resistant to multiple races of BLB but also much similar to the recurrent parent in terms of yield and grain quality traits. This effort clearly demonstrates the utility of MAS in backcross breeding programmes involving multiple/ recessive genes.

MAS and quantitative traits

Most traits of agronomic importance such as yield, abiotic stress tolerance etc. are complex in nature, controlled by several genes located on different chromosomes and significantly influenced by environmental conditions. Furthermore, the genes controlling such traits are also influenced by other genes located on different chromosomes (epistatic interaction). Compared to the traits controlled by major genes, improvement of quantitative traits thorough MAS raises more questions. Even though many QTLs spread across the rice genome are controlling a Quantitative trait (QT), only a few are detected in mapping studies largely because of epistatic and Genotype x Environment (GE) interactions. Hence the maker-QTL linkage inferences may also vary. There is also a major conceptual difference between molecular markers and their linked QTLs. Molecular markers mark QTL and not directly the quantitative trait. Molecular markers are analogous to Mendelian gene (ie., only diallelic single gene). On the other hand, QTLs are 'multigenic' with small 'major gene' effects and high epistatic interaction. Hence in a strict sense, QTLs have genotypes but no phenotypes. A lot more concerted effort is required to reliably map the QTL controlling complex traits. In mapping QTL, the percent phenotype variation under the control of each locus is generally used to assess the effect of specific locus on the trait. For any specific QTL, the percent of phenotype explained varies depending on the environmental conditions to which the population has been subjected. The most intriguing questions therefore often asked about QTLs are

   (i)   How often QTL uncovered in a mapping study is correct?

   (ii) How accurate are the estimates of genetic effects?

   (iii) Of all the relevant QTLs for a trait, how many have been actually discovered?

The difficulty in manipulating the quantitative traits is due to their complexity in number and interaction between the loci (i.e., epistasis). Since several genes located in different regions of the genome with small individual effects and their interaction are involved in the expression of the trait, several regions (QTL) are to be manipulated simultaneously. QTL for yield genes along with their associated markers have been identified in wild relatives of rice (*O. rufipogon*) by Xiao et al. 1998. This demonstrates another dimension of MAS, i.e., DNA marker technology might be useful for novel goals in rice breeding beyond performing the tasks of conventional breeding. A QTL present in a early generation segregant may go undetected in phenotypic selection because of the epistatic interaction involved and hence may be left unexploited, but a simple molecular marker analysis may reveal the existence of the QTL in the segregating progeny and hence may help in its targeted introgression. Similar ongoing work at DRR involved the introgression of yield contributing QTLs from land races and wild relatives of rice into popular parental line– KMR3 and a few elite varieties like IR64, Swarna, etc. Preliminary results clearly show that the introgressed elite varieties and hybrids developed possess a considerable yield advantage over the parental lines and checks.

Conclusions

An array of marker systems have been developed with a wide range of applications for gene tagging, linkage mapping, MAS, Map based cloning, QTL analysis etc. There are still only few reports of successful MAS performed in rice breeding. This could be attributed to difficulties in application due to high cost and lack of collaboration between molecular geneticists and breeders.

There is continuous requirement to saturate marker linkage maps with markers tightly linked to genes of interest. Marker systems need to be refined to make them more user friendly and cost effective. SSRs along with single nucleotide polymorphisms (SNPs) are expected to become important marker systems in the near future when the complete sequence databases become available. Micro array based markers (DART), Expressed sequence tag (EST) based markers, intragenic markers and development of allele specific markers coupled with advances in automation would certainly make MAS a much powerful and effective approach in crop breeding.

References

Ayers NM, McClung AM, Larking PD, Bligh HFJ, Jones CA, Park WD. (1997). Microsatellites and single-nucleotide polymorphism differentiate apparent amylose classes in an extended pedigree of US rice germ plasm. **Theor. Appl. Genet.** 94: 773-781.

Causse MA, Fulton TM, Cho YG, Ahn SN, Chunwongse J, Wu KS, Xiao JH, Yu ZH, Ronal PC, Harrington SE, Second G, McCouch SR, Tanksley SD. (1994). Saturated molecular map of the rice genome based on an interspecific backcross population. **Genetics** 138: 1251-1274

Dreher, K. et al. (2003). Money matters (I): costs of field and laboratory procedures associated with conventional and marker-assisted maize breeding at CIMMYT. **Molecular Breeding.** 11: 221-234

Hittalmani S, Parco A, Mew TV, Zeigler RS, Huang N. (2000). Fine mapping and DNA marker-assisted pyramiding of the three major genes for blast resistance in rice. **Theor. Appl. Genet.** 100: 1121-1128.

Huang N, Angeles ER, Domingo J, Magpantay G, Singh S, Zhang G, Kumaravadivel, N, Bennett J, Khush GS. (1997). Pyramiding of bacterial blight resistance genes in rice: marker-assisted selection using RFLP and PCR. **Theor. Appl. Genet.** 95: 313-320.

J. Yashitola, R. M. Sundaram, S.K Biradar, T. Thirumurugan, M. R. Vishnupriya, R. Rajeshwari, B. C. Viraktamath, N. P. Sarma and Ramesh V. Sonti. (2004). A sequence specific PCR marker for distinguishing wild abortive cytoplasm containing rice lines from their cognate Maintainer lines. **Crop Science** (*In press*).

Kurata N, Nagamura Y, Yamamoto K, Harushima Y, Sue N, Wu J, Antonio BA, Shomura A, Shimizu T, Lin SY, Inoue T, Fukada A, Shimano T, Kuboku Y, Toyoma T, Miyamoto Y, Kirihara T, Hayasaka K, Miyao A, Monna L, Zhong HS, Tamura Y, Wang ZX, Momma T, Umehara Y, Yano M, Sasaki T, Minobe Y. (1994). A 300-kilobase interval genetic map of rice including 883 expressed sequences. **Nat. Genet.** 8: 365-372

Lang NT, Subudhi PK, Virmani SS, Brar DS, Khush GS, LI ZK, Huang N. (1999). Development of PCR-based markers for thermosensitive genetic male sterility gene *tms3(t)* in rice (*Oryza sativa* L.). **Hereditas** 131: 121-127.

Mackill DJ, Ni J.(2001). Molecular mapping and marker assisted selection for major-gene traits in rice. *In:* Khush GS, Brar DS, Hardy B, editors. 2001. Rice genetics IV. Proceedings of the Fourth International Rice Genetics Symposium, 22-27 October 2000, Los Banos, Philippines, New Delhi (India): Science Publishers, Inc., and Los Banos (Philippines): International Rice Research Institute. Pp: 137-151.

McCouch SR, Kochert G, Yu ZH, Wang ZY, Khush GS, Coffman WR, Tanksley SD. (1988). Molecular mapping of rice chromosomes. **Theor. Appl. Genet.** 76: 815-829

McCouch SR, Teytelman L, Xu Y, Lobos KB, Clare K, Walton M, Fu B, Maghirang R, Li Z, Xing Y, Zhang Q, Kono I, Yaho M, Fjellstorm R, DeClerck G, Schneider D, Cartinhour S, Ware D, Stein L .(2002). Development and mapping of 2240 New SSR Markers for Rice (*Oryza sativa L.*). **DNA Research.** 9: 199-207.

Nair S, Bentur JS, Prasada Rao U, Mohan M (1995) DNA markers tightly linked to a gall midge resistance (*Gm2*) are potentially useful for marker-assisted selection in rice breeding. **Theor. Appl. Genet.** 91:68-73

Singh S, Sidhu JS, Huang N, Vikal Y, Li Z, Brar DS, Dhaliwal HS, Khush GS. (2001). Pyramiding three bacterial blight resistance genes (xa5, xa13 and Xa21) using marker-assisted selection into indica rice cultivar PR106. **Theor. Appl. Genet.** 102: 1011-1015.

Xiao J, Li J, Grandillo S, Ahn SN, Yuan L, Tanksley SD, McCouch SR. (1998), Identification of trait-improving quantitative trait loci alleles from a wild rice relative, *Oryza rufipogon*. **Genetics.** 92: 637-643.

Yashitola J, Thirumurugan T, Sundaram RM, Ramesha MS, Sarma NP and Sonti. RV. (2002). Assessment of purity of rice hybrids and parental lines using microsatellites and STS markers. **Crop Science.** 42: 1369-1373.

10

Molecular Markers as Tools for Genome Analysis : Assessment of Genetic Diversity in Select Medicinal Plants

M.S. Lakshmikumaran[1]*, A.Singh[1], M.S. Negi[1] and P.S. Srivastava[2]
Bioresources and Biotechnology Division, TERI, Darbari Seth Block, Habitat Place, Lodhi Road, New Delhi 110 003.
Faculty of Science, Centre for Biotechnology, Hamdard University, New Delhi 110 062.

India is home to a wide variety of genetic resources chiefly because of its diverse agroclimatic regions. The Eastern Himalayas and the Western Ghats represent two hotspots of the biological diversity, which is the reason why India has been identified as one of the 12-megadiversity centres of the world. India harbors atleast 45000 different plant species of which 15000 are angiosperms (WRI 1996, see Manoharan 1997). A sizable fraction of this flora is characterized as medicinal in nature and their pharamacokinetic properties are due to the presence of secondary metabolites such as, tannins, alkaloids, phenolics and terpenes that presumably play a role in the defense mechanism of plants against various types of stresses. Almost 80% people in the developing nations rely extensively on these medicinal plants for their primary health through traditional practices and tribal medicine. The traditional system of medicine such as Ayurveda, Siddha, Unani, Amchi and Tibetan have been very effective since time immemorial. More recently, these alternative forms of medicine have become extremely popular in Western countries because unlike allopathic formulations they are supposed to be devoid of side effects. Consequently, active research in medicinal plants has gained momentum and several modern pharmaceutical giants are bio-prospecting for active ingredients in the medicinal plants that can be developed as effective drugs or serve as models for synthesis of chemical analogs. This type of research is exemplified by plant based drugs that have met commercial success such as digitoxin, reserpine and tubocurarine. In fact, the export of herbal-based products from India is estimated to have reached US $ 1150 million by the end of year 2001 (Natesh 2000).

Unfortunately, only 10% of the hitherto identified medicinal plant species are cultivated in a sustainable manner. Most of these herbs are exploited from the wild and were gradually eliminated. Thus, a rapid loss of forest cover coupled with urbanization has caused erosion of our resource base. It is therefore vitally important to conserve our genetic wealth before it is lost irretrievably. Fortunately, in the post CBD era, the gene rich third world nations have gained sovereign rights to regulate the access to their germplasm and prevent bio-piracy. Guidelines have been defined pertaining to IPR issues for protection of the indigenous knowledge and bio-wealth. Therefore, it is not only important to conserve but also to protect our genetic wealth from exploitation.

Systematic documentation database of plant species through assessment and characterization of genetic diversity is an essential first step towards this end. Also, the role of biotechnology in addressing this issue cannot be ruled out (Westman and Kresowich 1997). The role of biotechnology in germplasm cataloguing and effective utilization has been categorized by Barlow and Tzotos (1995) as :

1. Providing conservation technologies such as tissue-culture, seed banks and gene banks.

2. Effective management of germplasm collection (*in-situ/ex-situ*) through :
    · Monitoring and assessing genetic diversity and allowing precise analysis of the existing diversity.
    · Formulating critical conservation strategies on what should be conserved on a priority basis, which involves determining the number of genotypes that might represent the entire or near-entire diversity. These genotypes at a particular level of taxon may represent a core collection.
    · Evaluating genetic relationships between different taxa.

3. Improvement of existing germplasm by:
    · Marker assisted breeding and selection,
    · Genetic manipulation by transferring important genes thereby broadening the genetic base.
    · Identifying genotypes that are of immediate practical value or hold promise for future exploitations.

Genetic markers serve as efficient tools for wide-scale analysis of plant genetic resources. They are defined as landmarks on the genome, since they refer to specific locations, that may be appropriately mapped on the genome. A wide array of molecular genetic markers (classical as well as modern) find applications in genome analysis depending upon the objective and the availability of the resources. However, few desirable features make one class of molecular marker preferable over the other. These are:

- Detection of polymorphism to allow discrimination of genotypes.
- Co-dominance permitting discrimination of homozygous from the heterozygous state. This allows localization of DNA sequences at a definite locus on the genome which is of utmost importance in mapping programs.
- Unaffected by environmental factors and developmental stage of the plants.
- Should not be influenced by epistatic interactions and pleiotropic effects.
- Preferably uniformly distributed to ensure wide genome coverage.
- Efficient, fast, reproducible and economical.

The classical strategies employed for analyzing genetic variability relied solely on the visual inspection through comparison of phenotypic traits such as morphology, anatomy and embryology. Subsequently, protein based markers (isozymes and allozymes) were introduced wherein, polymorphism in the electrophoretic migration of proteins were utilized. However, protein based markers are an outcome of gene-expression and provide us only an indirect way of visualizing variation at the DNA level. These markers are under the control of environmental cues, developmental stage and are influenced by pleiotropic effects and epistatic interactions. Thus, inferences based on such markers are at times anomalous.

DNA Based Markers

The introduction of DNA based markers circumvented the limitations encountered by morphological and protein based markers and completely revolutionized the area of molecular genetic analysis. DNA based markers scan the genome directly and are, therefore, neutral to environmental factors, developmental stage and are not influenced by other genes and factors. DNA based markers not only rely on direct evaluation of genome but also provide an abundant and unlimited supply of markers. Moreover, the variations at the DNA level are far more frequent in comparison to the variability at protein level. Some of the desirable traits presented by ideal DNA based markers include polymorphic nature, co-dominance, high multiplex ratio, wide-spread occurrence to ensure a thorough genome coverage, simple and cost-effective, amenable for automation and high reproducibility and reliability. DNA based markers have been classified into hybridization and PCR/ Amplification based markers.

Hybridisation Based Markers

Hybridisation based markers detect polymorphisms caused either at, or in regions lanked by restriction sites. Such markers are exemplified by RFLP and satellite (mini - and microsatellite) DNA probes. Procedurally, all the hybridisation based marker systems involve restriction of genomic DNA followed by electrophoresis, blotting, hybridisation and detection. However, there are variations to the set procedures, for instance, in-gel hybridisation is preferred when using oligonucleotides as hybridisation probes. The hybridisation-markers are sub-divided into two sub-classes employing probes that target either single locus or multiple-loci.

Restriction Fragment Length Polymorphism

RFLP markers were first developed for human genome analysis (Botstein et al. 1980) and then adapted for plant genomes. DNA nucleotide variation resulting from point mutations occasionally leads to loss or gain of restriction endonuclease site. DNA alterations involving larger regions can be a result of insertion, deletion, inversion and/or translocation events. Both the causes mentioned can result in re-distribution of several restriction endonuclease sites. The size distribution of DNA fragments affected by such changes will be different from that of an unaffected individual thus resulting in a RFLP pattern. Single or low copy sequences in the genome are ideal RFLP probes and they survey the coding region. This makes single copy RFLP suitable for genome and synteny mapping (Lespinasse et al. 2000; Cheung et al. 1997), phylogenetic analysis (Zheng et al. 1994; Bukhari et al. 2000) though some authors have employed them for genetic diversity analysis (Helentjaris et al. 1985; Garcia Mas et al. 2000). RFLP requires use of a large amount of highly pure DNA, coupled with radioactivity for detection purposes. The steps involved are lengthy and laborious which act as a deterrent towards its acceptance for large-scale application. A high degree of reliability is the hallmark of RFLP markers.

Mini- and Microsatellite and Oligonucleotide markers

Satellite DNA sequences are high copy and are present as tandem repeats. The organisation of satellite DNA is quite complex as these (mini and microsatellites) are dispersed at many loci and also interspersed among themselves and other repetitive elements. Simple short motifs arranged in a tandem array of di-, tri-, tetra- or penta-nucleotide units are called microsatellites or simple sequence repeats (SSR). Microsatellites may be classified into simple or compound core units. Simple microsatellites are exemplified by such dimeric, trimeric or tetrameric core units such as $(CA)_n$, $(CAA)_n$, $(GATA)_n$ repeated 'n' number of times. In some instances, particularly plant genomes, it has been shown that two different SSR units are located adjacent to each other to give rise to a compound microsatellites (Vogel and Scolnick 1998). These have structures like $(GATA)_n GT(CAC)_n$,

$(CT)_n(GT)_n$, $(ATA)_nGCC(TAT)_n$. The ubiquitous nature of microsatellites has made it feasible to use them for analysis of genetic diversity in a number of plants (Parasnis et al. 1999; Assefa et al. 1999; Perrera et al. 2000; Testolin et al. 2000).

Minisatellites are tandem repeats where the motifs comprise of larger monomeric units ranging from 10 to 60 bp. Minisatellite sequences were isolated from the intron region of the human myoglobin gene and termed as Jefferey's minisatellite, 33.15 and 33.6 (Jeffrey et al. 1985). Similar minisatellites are not very frequently found in plant genomes though there are a few reports of their isolation and characterisation (Broun and Tanksley 1993) and use in genetic diversity analysis (Dallas 1988).

Synthetic oligonucleotides such as $(GATA)_4$ and $(GACA)_4$ have been routinely used as probes for DNA fingerprinting of plants, variety identification and hybrid analysis (Weising et al. 1989; Beyermann et. al 1992; Bhatia et al. 1995; Blair et al. 1999; Perrera et al. 2000; Prasad et al. 2000; Testolin et al. 2000) and genome mapping and tagging (McCouch et al. 1997; Parasnis et al. 1999; Peng et al. 1999; Cervera et al. 2001).

Satellite DNA probes detect polymorphism on several accounts. Firstly, a feature commonly observed in all repetitive sequence arranged in tandem is their high degree of polymorphism when compared between genomes. Such polymorphisms are due to the variation in the number of repeats due to events like unequal crossing-over, insertions, deletions and amplifications (Charlesworth et al. 1994). On such account, the alternative name for these hypervariable satellite regions is Variable Number of Tandem Repeats (VNTR). This VNTR feature is a major source of polymorphism. Secondly, satellite units are known to be located within other repeats thus making these 'other' repetitive sequences a source of hypervariability. Finally, the variations at the restriction sites are additional source of polymorphism at the mini - and microsatellite loci. Due to their capability to generate high polymorphism, satellite DNA probes are highly informative and enable the generation of unique DNA fingerprints. The repetitive nature of satellite DNA enables them to hybridise to multiple loci and generate dominant markers. Such satellite DNA probes, particularly microsatellites, were initially developed as hybridisation based markers, and later have been combined with the PCR technology to develop far more efficient marker systems commonly called as MP-PCR.

PCR based Markers

The extremely powerful and sensitive technique of PCR (Mullis et al. 1986) paved the way for the development of a variety of PCR-based molecular markers. The emphasis shifted from use of cumbersome, lengthy hybridisation based markers to technically much simplified PCR based markers. Principally, all PCR based markers detect changes that affect a primer binding site or the distribution of primer binding sites. The polymorphism at the primer-binding site is due to sequence variation, and the variation internal to the primer's annealing sites get reflected as differences in the size of amplification products. Though initially developed to target a particular locus, several modifications allowed PCR to analyse anonymous regions based on arbitrary and semi-arbitrary primers. PCR has been put to a variety of uses after suitable improvisations, depending on the type of primer used. A target PCR uses a primer pair that amplifies a defined region and generates co-dominant marker. It is thus suited for mapping and tagging of genes using Bulked Segregant Analysis (BSA), Expressed Sequenced Tags (EST), Sequence Tagged Sites (STS), Cleaved Amplified Polymorphic Sequences (CAPS) and Sequence Characterised Amplified Regions (SCAR). On the other hand, arbitrary or semi-arbitrary primed PCR techniques are multi-allelic and generate dominant products. These, therefore, are applied for genetic diversity assessment through DNA fingerprinting using Random Amplification of Polymorphic DNA (RAPD) and Amplified Fragment Length Polymorphism (AFLP).

Target PCR

In a typical target PCR, prior sequence information of the locus is required in order to synthesise site-specific oligonucleotide primers. These primers anneal to the complimentary sequences on the single-stranded template DNA, on either side of the target locus and cause amplification. The amplification products correspond to same locus from various individuals, and thus are allelic or co-dominant. Some of the target PCR markers are CAPS (Konieckzny et al. 1993), EST (Adams et al. 1991), SCAR (Paran and Michelmore 1993) and STS (Thomas and Scott 1994). These markers have applications in the field of genome mapping.

CAPS/EST/STS

The molecular investigations are a source of sequence information to develop a variety of target PCR markers. Sequencing of cDNA clones from both ends generate Expressed Sequence Tags (Adams et al. 1991). An EST can therefore be used to design primers that would be specific to a particular locus. Similar locus-specific primer pairs can also be designed based on other regions. For example, co-dominant RFLP probes, and dominant markers like RAPD and AFLP can be sequenced. This sequence information can be converted into target specific primer pairs like Sequence Tagged Sites (STS; Thomas and Scott 1994) and Sequence Characterised Amplified Regions (SCAR; Paran and Michelmore 1993). The locus-specific primers, whether EST, STS or SCAR may not detect polymorphism between individuals through simple PCR. In such instances, the amplified product is restricted to screen for polymorphism at the restriction enzyme recognition sequences present internal to the primer annealing sites. Such an approach involving amplification followed by restriction led to the development of Cleaved Amplified Polymorphic Sequences (CAPS; Konieczny and Ausubel 1993) CAPS marker can be used across individuals of a segregating population so that the resultant polymorphism scored can then be mapped on to a linkage group. The primers designed for markers like CAPS, EST, SCAR, and STS based on the sequence data of one taxa may not be applicable to another taxa due to specificity of the sequence except when the primers target a region that shows high degree of sequence conservation across genera.

Arbitrarily-Primed and Semi-Arbitrarily-Primed or Anchored PCR

As the term suggests, primers are not specific to a locus and amplify arbitrary or anonymous sequences. Therefore, no prior information about the genome is required and a set of randomly generated primers can be used across number of different taxa. Since the primers are arbitrary, they amplify multiple loci and the markers are dominant. The arbitrary primed-PCR markers are highly discriminatory and most of the applications are related to DNA-fingerprinting and genetic diversity analysis. However, with proper experimental strategy, arbitrary and semi-arbitrary primed PCR can also be applied for mapping activities. Several markers based on arbitrary or semi-arbitrary primed PCR have been developed like RAPD, AFLP and SAMPL.

Randomly Amplified Polymorphic DNA

The first arbitrarily primed-PCR techniques were simultaneously developed by two independent groups (Welsh and McClelland 1990; Williams et al. 1990). The two nearly identical techniques were termed as Random Amplification of Polymorphic DNA (RAPD; Williams et al. 1990) and Arbitrarily Primed-PCR (AP-PCR; Welsh and McClelland 1990, 1991). The technique of RAPD involves PCR of genomic DNA with single arbitrary, short primers (usually decamer primers). Use of such primers and low annealing temperatures (37°C to 40°C) ensures several complementary sites distributed randomly in the genome to bind to random primer and give rise to amplification products. After the PCR, the amplification products are electrophoresed and visualised by ethidium bromide staining. Each amplification product is derived from a region of the genome that contains DNA segments with homology to the primer; these segments should be present on the opposite strands of DNA and be sufficiently close (500bp to 2500bp) for amplification. The polymorphisms detected are a result of either a nucleotide change that alters the primer-binding site, or of an insertion/deletion within the amplifiable region. Polymorphisms at different loci are therefore scored as presence or absence of a particular band. RAPD markers, thus, are dominant in nature and alleles for a particular locus cannot be differentiated.

RAPD is a method of choice for assessing genetic variability, estimating genetic relatedness and fingerprinting of plant genomes (Russell et al. 1997;

Barker et al. 1999; Rodriguez et al. 1999; Rajgopal et al. 2000; Das et al. 1999). The technique gained importance due to its simplicity, efficiency and non-requirement for prior sequence information. Each reaction would normally amplify 10-15 scorable bands. Although popular, several authors have reported non-reproducibility as a major drawback associated with RAPD technique (Davin-Regli et al. 1995; Jones et al. 1997). RAPD has also been employed for genetic mapping and tagging of genes following Bulked Segregant Analysis (BSA; Michelmore et al 1991; Bannerjee et al. 1999).

Amplified Fragment Length Polymorphism

The AFLP technique developed by Vos et al. (1995) and Zabeau and Vos (1993) is now a preferred tool for various genetic analyses. These include diversity assessment and construction and saturation of linkage maps. The wide acceptance and applicability of AFLP is because it combines the best features of RFLP and PCR i.e. reliability, reproducibility derived from RFLP and sensitivity of PCR. AFLP routinely generates nearly 50-100 detectable products during each assay. These features are a quantum jump over features associated with other previously developed marker systems. Principally, AFLP is based on selective amplification of a subset of restricted DNA fragments using the PCR technique. Procedurally, the DNA is simultaneously restricted with two enzymes, one a frequent cutter (such as *Mse* I) and a rare cutter (such as *Eco* RI). An alternative combination that is suggested is *Taq* I + *P.st* I. The restricted fragments are then ligated to enzyme-specific adapters to generate the template for PCR. The collection of template after adapter-ligation generates fragments with different length and sequence but with identical ends (due to the adapters used). The primer pair (one for *Eco* RI and one for *Mse* I adapter) that are employed for PCR are thus anchored to the common adapter sequences. However, the primers contain selective nucleotide at their 3' end that extends into the template DNA fragment. This anchored, semi-arbitrary primer selectively amplifies a subset of the huge population of template that is present in the AFLP library. The number of amplification products can be tuned by adjusting the number and choice of selective nucleotide at the 3' end of the primer. The primers are radioactively labelled to facilitate autoradiography. Each experiment, on an average yields about 50-100 bands in the size range of 50-350bp. The fragments are electrophoresed on denaturing polyacrylamide gel. AFLP has found widespread application in assessment of the genetic diversity studies (Singh et al.2002; Hongtrakul et al. 1997; Amsellem et al. 2000), for construction and saturation of linkage maps, and tagging of genes (Becker et al. 1995; Voorrips et al. 1997; Terauchi and Kahl. 1999).

Microsatellite Primed PCR (MP-PCR)

This terminology is commonly applied to a set of techniques which involve SSR based sequences to prime a PCR reaction, essentially to detect polymorphism. These have been exemplified by techniques such as SAMPL, multi-loci profiling techniques including Inter-SSR Amplification (ISA/ISSR), Randomly Amplified Microsatellite Polymorphism (RAMP).

Selectively Amplified Microsatellite Polymorphic Loci

Morgante and Vogel (1994) developed a modified version of the AFLP technique termed as Selective Amplification of Microsatellite Polymorphic Loci (SAMPL). This strategy targets the ubiquitous microsatellite loci by the use of compound microsatellite primer in conjunction with an AFLP primer during the selective amplification stage of the AFLP library. The use of a compound microsatellite primer in SAMPL is dictated by the fact that such compound satellites are commonly found in plant genomes (Vogel and Scolnick, 1998). The use of simple microsatellite primers may not produce distinct products. This is because the simple 'unanchored' primers anneal to different targets at the same locus and results in a mixture of heterogeneous products. This leads to smearing during electrophoretic resolution. However, compound microsatellite acts as an anchored primer, leading to specific binding and amplification of discrete products. As the frequency and type of microsatellites varies in different organisms, there is a need to screen a large number of SAMPL primers and select the ones giving optimal amplification products. As the microsatellite loci are hypervariable, SAMPL markets are likely to be useful in studies where low genetic diversity is expected.

To summarise, a marker that exhibits co-dominance is usually suitable for mapping purposes. These include RFLP, CAPS, STS (Jin et al. 2000, Caranta et al. 1999, colombani et al. 2000) etc. Dominant markers that are multiallelic can be very effectively utilised for genetic diversity assessment. These markers include microsatellites, RAPD, AFLP and SAMPL (Elias et al. 2000; Fregne et al. 2000; Roa et al. 2000, Singh et al. 1999, 2001). However, the compartmentalisation of markers and their applications based on dominance/ co-dominance is not watertight. For example, co-dominant markers have been used for genetic diversity and phylogenetic studies (Bukhari et al. 1999; Garcia-Mas et al. 2000), and dominant markers have been used in mapping and tagging of traits (Parasnis et al. 1999; Terauchi and Kahl 1999). A comprehensive list of plant species and various markers used to study genetic diversity, construction of linkage maps and tagging of traits have been provided in Table 10.1.

Table 10.1 List of plant species, markers used and their utility

Plant Species	Marker	Uses	Reference
Oryza species	AFLP; Microsatellite	Phylogenetic analysis; Linkage mapping	Aggarwal et al. 1999; Mackill et al. 1996; McCouch et al. 1997
Rubus alceifolius	AFLP	Genetic diversity	Amsellem et al 2000
Olea spp. Olea europea, O. laperrini, O. maroccana, O. chrsophylla, O. ferruginea, O. africana	AFLP	Genetic diversity	Angiolillio et al. 1999
Cynodon	DAF	Genetic diversity	Assefa et al. 1999
Salix spp.	RAPD and AFLP	Genetic diversity	Barker et al. 1999
Arabidopsis thaliana	AFLP	Genetic diversity	Breyne et al. 1999 Erschadi et al. 2000
Acacia	Chloroplast RFLP	Phylogenetic analysis	Bukhari et al. 1999
Lolium spp. (rye grass)	AFLP	Genetic diversity	Cresswall et al. 2001; Roldan-Ruiz et al. 2000
Rhododendron simsii (Evergreen Azaleas)	AFLP	Genetic variation	De Rick et al. 1999
Marnihot esculenta	AFLP and microsatellite	Genetic variation and germplasm chracterization for disease resistance	Elias et al. 2000; Fregne et al. 2000; Roa et al. 2000
Pisum spp. (Pea)	Ty-1 copia class retrotransposon	Diversity analysis and mapping	Ellis et al. 1998
Melon	AFLP, RAPD and RFLP	Genetic diversity	Garcia-Mas et al. 2000
Eryngium alpinum	AFLP	Genetic diversity of the endangered plant	Gaudel et al. 2000
Alstroemeria species	AFLP	Genetic diversity in Chilean and Brazilian species	Han et al 2000
Chichorium species	AFLP	Diagnostic marker for endive and chicory group	Kiers et al. 2000
Musa spp., Musa acuminata	AFLP, RAPD	DNA fingerprinting and Genetic diversity	Loh et al. 2000; Pillay et al. 2001
Datura spp.	AFLP	Genetic diversity	Mace et al. 1999
Moringa oleifera	AFLP	Genetic variation	Muluvi et al. 1999
Digitalis obscura L. (Willow leaved foxglove)	RAPD	Genetic variation within and between natural population	Nebauer et al. 1999

Table 10.1 Contd....

Plant Species	Marker	Uses	Reference
Withania somnifera and *W. coagulans*	AFLP	Genetic variation within and between species	Negi et al. 2000
Papaya	Microsatellite (GATA)$_4$	Sex-specific differences	Parasnis et al. 1999
Cocus nucifera (coconut)	Microsatellite	Genetic diversity and population studies	Perrera et al. 2000
Capsicum sps.	RAPD	Genetic variation	Rodriguez et al. 1999
Pedicularis palustris	AFLP	Genetic variation in the remnant population of this rare plant and correlation to the population size and reproduction	Schmidt and Jensen 2000
Morus spp.	AFLP	Genetic diversity	Sharma et al. 2000
Sasa senarensis (Dwarf bamboo)	AFLP	Clonal structure	Suyama et al. 2000
Dioscorea tokoro	AFLP	Mapping and linkage of AFLP marker to sex	Terauchi and Kahl 1999
Prunus persica	Microsatellite	DNA fingerprinting and genetic origin of cultivars	Testolin et al. 2000
Astragalus cremnophylax var. cremnophylax	AFLP	Genetic variation in this critically endangered plant species	Travis et al. 1996
Cryptomeria japonica	STS	Genetic diversity	Tsumura and Tomaru 1999
Taxarum officinale (Triploid Dandelions)	AFLP	Genetic diversity and population structure	Van der Hulst et al. 2000
Stylosanthes sps.	STS	Genetic relationship	Vander Stapper et al. 1999
Mangifera sps.	AFLP	Genetic relationship	Eiadthong et al. 2000
Piper longum	RAPD	Development of male-sex associated marker	Banerjee et al. 1999
Barley	RFLP, AFLP, SSR, RAPD	Genetic variation	Russell et al 1997
Azadirachta indica and *A. indica* var. siamensis (Indian and Thai neem)	AFLP and SAMPL	Genetic diversity at inter- and intra-population level	Singh et al. 1999, 2001
Castanea sativa Mill. (European chestnut)	RAPD, ISSR, Isozymes	Genetic linkage map	Casaoli et al. 2001
Phyllanthus emblica	Isozyme	Genetic diversity	Uma Shanker and Ganeshiah 1997

Application of DNA based Markers for Germplasm Characterisation

The characterisation of germplasm and the estimation of inherent genetic diversity plays an important role in germplasm identification, cataloguing and management, and in formulating conservation policies. In this article we present two case studies on the characterisation and analysis of diversity in two Indian medicinal plants namely Ashwagandha (*Withania somnifera*) and Amla (*Phyllanthus emblica*).

Case study I: Analysis of genetic diversity in Ashwagandha (*Withania somnifera* and *W. coagulans*)

Withania somnifera, member of the Solanaceaen family is known for its medicinal properties, which are attributed to several withanolides, and other active compounds like steroidal lactones (withaferins) and alkaloids (Srivastava et al. 1960; Uma Devi 1993, 1996). *Withania* species grows in the dry parts of sub-tropical India. Atal and Schwarting (1962) reported that there exists a very high degree of variability in morphological features and growth habits not only among *Withania* that grows in India but from other countries. Inspite of the variation that exists, there has been no attempt to classify the different morphotypes of *Withania* species, and the attempt to assess the genetic diversity reported by the authors (Negi et al. 2000) is the first of its kind in the genus *Withania*.

In the study reported by Negi et al. (2000), two distinct morphotypes of *W. somnifera* namely, *W. somnifera* var. Nagori and var. Kashmiri were analysed for variation using AFLP. Five individuals of the species *W. coagulans*, which grows as a xerophyte in Baluchistan, were also analysed. The study was conducted using AFLP markers to estimate the genetic similarity and the relationship within and among *Withania* species. Seven selective primer combinations were used which generated a total of 520 amplification products with an average of 74 bands per assay. Among the *W. somnifera* genotypes, 52 % of the bands were polymorphic; however the value went up to nearly 82% when *W. coagulans* genotypes were also included. The number of bands generated by each primer combination ranged from 53 in $E_{ACT} \times M_{CAT}$ to 101 in $E_{AAC} \times M_{CAA}$. Figure 10.1 shows a representative fingerprint that was obtained using primer combination $E_{ACC} \times M_{CAT}$. This fingerprint included genotypes of both *W. somnifera* species (var. Kashmiri and Nagori, lanes A-I and P-T, respectively) and *W. coagulans* (lanes U-Y). Ninety bands were scored across the genotypes of which 77 % were polymorphic between *W. somnifera* and *W. coagulans*. Some bands were present across all the *Withania* genotypes, irrespective of the species whereas some bands could clearly distinguish between the two species such as those that are present in all *W. somnifera* genotypes and absent from W. coagulans. The banding profiles of the species were distinct and the W. coagulans individuals could be easily differentiated from *W. somnifera* individuals. Similarly, the two distinct ecotypes of *W. somnifera* namely Kashmiri and Nagori were different at the genome level.

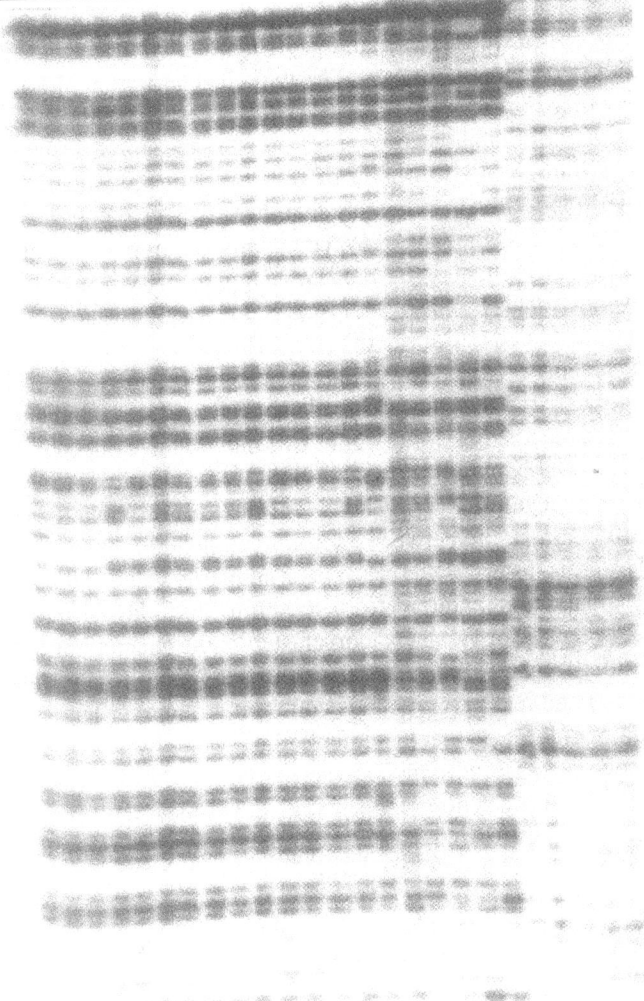

Kashmiri	I	Nagori	W.c
A B C D E F G H I J K L M N	O	P Q R S T	U V W X Y

I   :   Intermediate

W.c : *Withania coagulans*

Fig. 10.1 A representative AFLP fingerprint obtained with primer combination E_{ACC} X M_{CAT} across 25 *Withania* accessions. The lane marked O represent a variant which is intermediate between a Kashmiri and Nagori morphotype.

The seven AFLP primer combinations employed generated 520 amplification products across the 35 *W. somnifera* and five *W. coagulans* genotypes. The binary matrix was used for calculating the similarity values using Jaccard's coefficient (Jaccard 1908). A phenogram based on UPGMA analysis (Sneath and Sokal 1973) was constructed. This dendrogram depicting phenetic relationships among and within *Withania* species clearly separated *Withania somnifera* and *W. coagulans* individuals into two distinct clusters. The two clusters joined at a low similarity value of nearly 0.3. Further sub-clustering was observed within these two major clusters. The five individuals of *W. coagulans* were organised in two sub-clusters that joined at a high similarity value of 0.99. Within this subcluster, ties were observed between individuals indicating that these may be clones derived from the same mother plants. The analysis of W. coagulans genotypes indicated low levels of diversity. The narrow genetic base of W. coagulans may be ascribed to the fact that only a few individuals had been analysed. Moreover, these have been introduced recently from Baluchistan to Barmer district in Rajasthan. Among the *W. somnifera* genotypes, the Nagori variety clearly segregated out from the Kashmiri variety and within the Nagori cluster, the values ranged between 0.98 to 1.0 indicating a narrow genetic base. This may be possible as only five individuals collected from Udaipur (Rajasthan, India) were analysed. The individuals of Kashmiri variety also formed four sub-clusters with values ranging between 0.94 to 1.0 again indicating a narrow genetic base. In contrast, the between-morphotype diversity was quite high as the Kashmiri and Nagori clusters joined at a genetic similarity value of 0.51. Interestingly, one genotype clustered separately from Kashmiri and Nagori morphotypes. Its position was between Kashmiri and Nagori clusters in the phenogram. This individual also had a distinct fingerprint pattern, as can be seen from figure (lane o, marked as intermediate-I) and had several bands common with either Kashmiri or Nagori morphotypes. It was inferred that this intermediate was a hybrid between Kashmiri and Nagori genotypes. The intermediate separated out during cluster analysis but was closer to Kashmiri morphotypes as revealed by the genetic similarity value of 0:8 joining the intermediate and the Kashmiri group, and its morphology (Negi et al. 2000).

Case study II : Analysis of genetic diversity in *Phyllanthus emblica* accessions collected from different parts of India

The medicinal properties of amla have been popular in Asian countries, particularly the Indian sub-continent. Amla is known for its nutritional properties and is an important dietary source of vitamin C, minerals and amino acids. Amla is used as an antiscorbutic and in treatment of several digestive disorders. The fermented juice of amla is prescribed in jaundice, dyspepsia and cough. The fruit has diuretic properties and acts as a potent laxative. The active ingredients of amla reside in the fruit and are characterized as a mixture of 13 different tannins and four colloidal complexes including phyllaemblic (Zhang and Tanaka 2000). Amla has also been implicated for its anti-stress (Bhutada 1999), anti-fungal (Mehmood and Ahmad 1999) and anti-bacterial (Akhtar and Rahber 1997) properties. Atleast one US patent has been filed on amla by Surendra Rohatgi from Kanpur (US patent no. 55 29779). The patent is being defined as Ayurvedic composition for the prophylaxis and treatment of AIDS, flu, TB and other immuno-deficiencies. However, there have been no systematic efforts for the conservation of genetic resources, and also no attempt has been made hitherto to analyze the genetic diversity in amla using DNA based markers. This study was hence undertaken with an objective of gaining a preliminary insight into the nature and extent of genetic diversity by using AFLP markers in amla accessions collected from different parts of India.

The levels of genetic diversity were analysed in 30 accessions of Amla collected randomly from different parts of India. Sixteen amla accessions (A to P) were obtained from Medicinal Plant Conservation Areas (MPCA, Bangalore). These accessions were originally collected from BRT hills (Western Ghats). This area has been identified for *in-situ* conservation by Foundation for Revitalization of Local Health Tradition (FRLHT), Bangalore. Four elite genotypes (Q to T) were collected from amla farms in Faizabad (U.P.). Of these, Q, R and S belonged to the Banarasi variety while the accession named as T has been identified as Francis variety. Accessions U, V, W and X were collected from Haryana (Gual Pahari), Accessions Y, Z, A1 and A2, A3 and A4 from diffrent parts of Delhi (Hamdard University Campus and University of Delhi, South Campus). Map of India representing the sites of collection is provided in Fig. 10.2.

AFLP Analysis

A total of six primer combinations were employed to generate a total of 333 scorable AFLP bands across 30 amla accessions. Table 10.2 provides the details of the information conveyed by different primer combinations. On an average, 55.5 bands were amplified per assay of which 62% were polymorphic in nature. The highest

Table 10.2 Informativeness of various AFLP primer combinations employed during diversity analysis in amla

Primer Combination	Total number of bands	Total no. of polymorphic bands	Total no. of monomorphic bands	Percentage polymorphism
E-ACA X M-CTA	74	55	19	74
E-AAG X M-CTG	65	37	28	57
E-ACA X M-CAG	62	45	17	73
E-ACG X M-CTA	46	28	18	61
E-ACG X M-CTG	32	16	16	50
E-AAG X M-CAT	54	31	23	57
	Total = 333 Av. = 55.5	Total = 212 Av. = 35.3	Total = 121 Av. = 20	Av. = 62

number of amplification products (74) were obtained with primer combination E-ACA X M-CTA. Of the 74 bands scored, only nineteen AFLP products were monomorphic. Hence, this primer combination detected 74% polymorphism making it the most efficient of all the six primer combinations screened. The least number of scorable bands (32) were detected with primer combination E-ACG X M-CTG. Interestingly, with this primer combination, exactly 50% bands were polymorphic.

Sites for amla sample collection

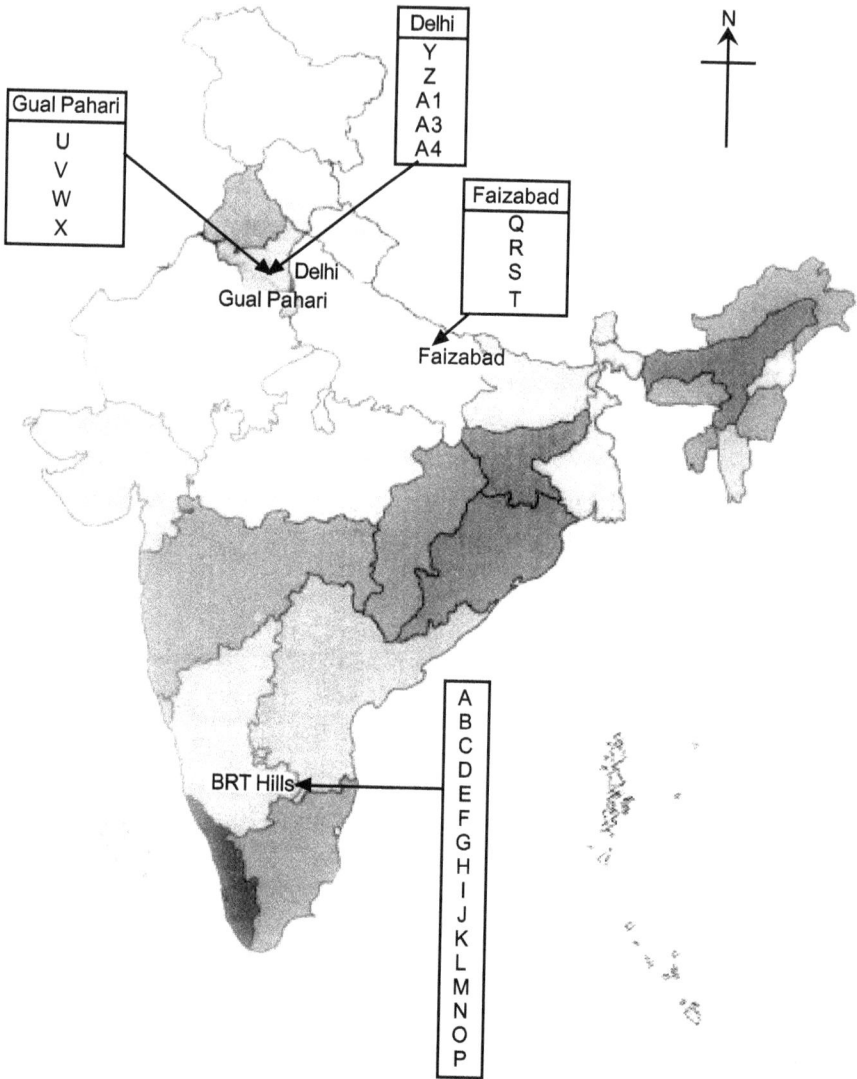

Fig. 10.2 Geographical localisation and regions from where amla accessions were obtained. The letters represent the name of the amla accessions.

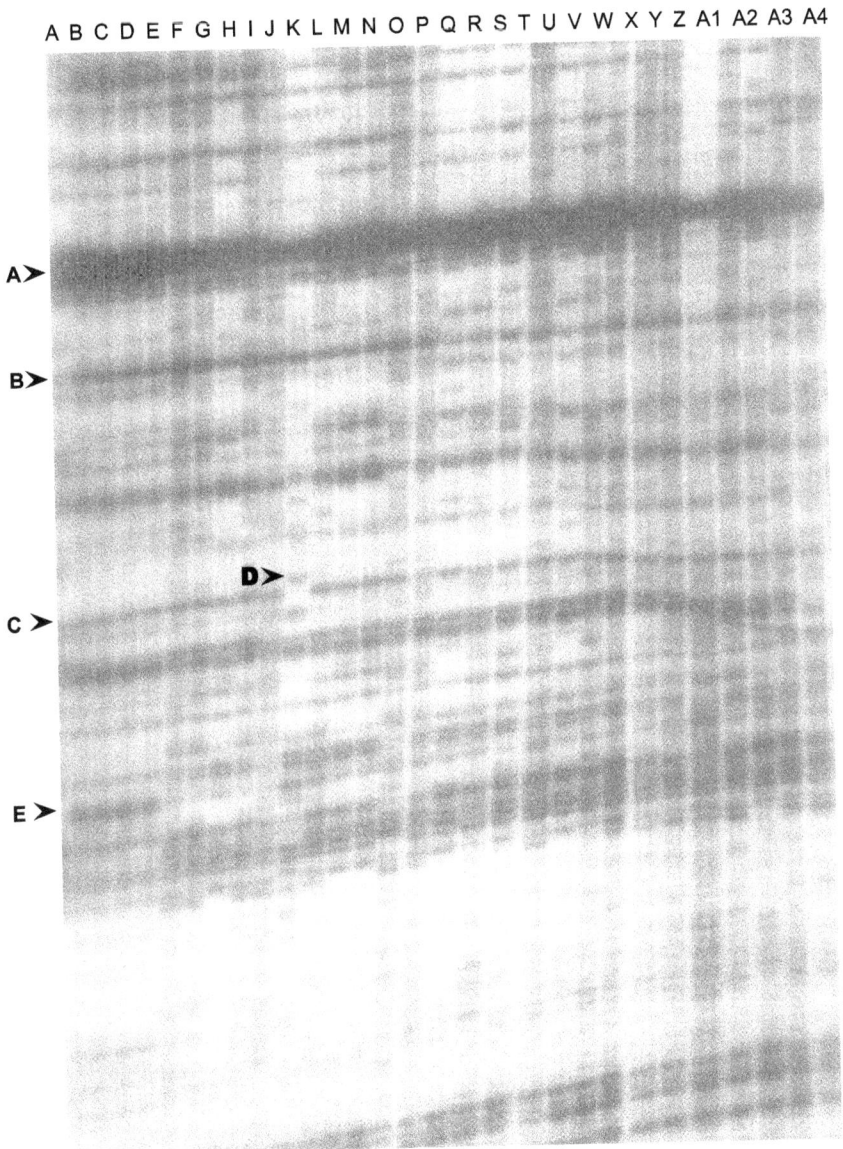

Fig. 10.3 AFLP fingerprint generated on employing primer combination E-ACA X M-CTA across the 30 amla accessions. AFLP fragments marked A and B represent monomorphic bands. Band marked D indicates polymorphic band which is rare unlike band C which is frequent. AFLP fragment marked E is another frequent polymorphic band.

Figure 10.3 is a characteristic AFLP profile generated on employing primer combination E-ACA X M-CT A. As discussed above, this primer combination was most efficient in generating highest number of bands and detecting highest percent polymorphism (see Table 10.2). As is clear in the fingerprint, a large number of fragments were detected thereby indicating that amla has a complex genome. Amplification products A and B represent monomorphic bands since these were detected in all the genotypes analyzed. Monomorphic band, A is especially interesting because of its strong intensity indicating that such bands are of repeat DNA origin. B and marked D represent a polymorphic fragment which is rare in the amla germplasm. On the contrary, band E is a frequent polymorphic band. The lane labeled as K (corresponding to the accession from Bangalore) displayed a unique fingerprint, which was different from the rest of the genotypes. For instance, band C was exclusively absent from this individual but was detected in all the other genotypes.

Figure 10.4 is another AFLP fingerprint profile obtained on employing primer combination E-AAG X M-CAT. Although a large number of AFLP fragments could be observed, only 54 fragments were unambiguous and could be scored reliably. The rest were ambiguous in nature and were not included in the binary matrix. Of the 54 AFLP fragments, 57% (corresponding to 31 bands) were polymorphic in nature. The remaining 23 bands displayed monomorphism. AFLP fragment marked A (arrow) was amplified across all the 30 accessions. Thus, this fragment represents a monomorphic band. On the other hand, polymorphic amplification product marked as B was amplified in some accessions but absent from few indicating that it is frequent in its occurrence. AFLP fragment marked with arrow C exemplifies a rare polymorphic AFLP fragment as it was detected only in accession K.

Statistical data analysis

Bands were scored for their presence (1) and absence (0) across the 30 accessions evaluated. The binary matrix thus generated was used to catculate a similarity matrix based on Jaccard's coefficient (data not shown). The average similarity value shared by 435 accession pairs was 0.77 indicating that our germplasm collection represented a broad genetic base with a large amount of genetic variation. The highest similarity value of perfect 1.00 was shared by accessions C and D, collected from Bangalore. In other words, these accessions were genetically identical and the six primer combinations employed could not differentiate the two accessions. Similarly, accession pairs B, C and B, D shared a very high similarity value of 0.99 each. However, these accession pairs were not identical. This indicates that individual B is close to both accessions C and D. The most dissimilar genotypes identified are within amla accessions K and N, sharing a similarity coefficient value of 0.57. A close examination of similarity matrix revealed that both the accessions K and O shared lower than average similarity values with all the other accessions. Hence, it is suggested that both these accessions are genetically distinct from rest of the amla accessions.

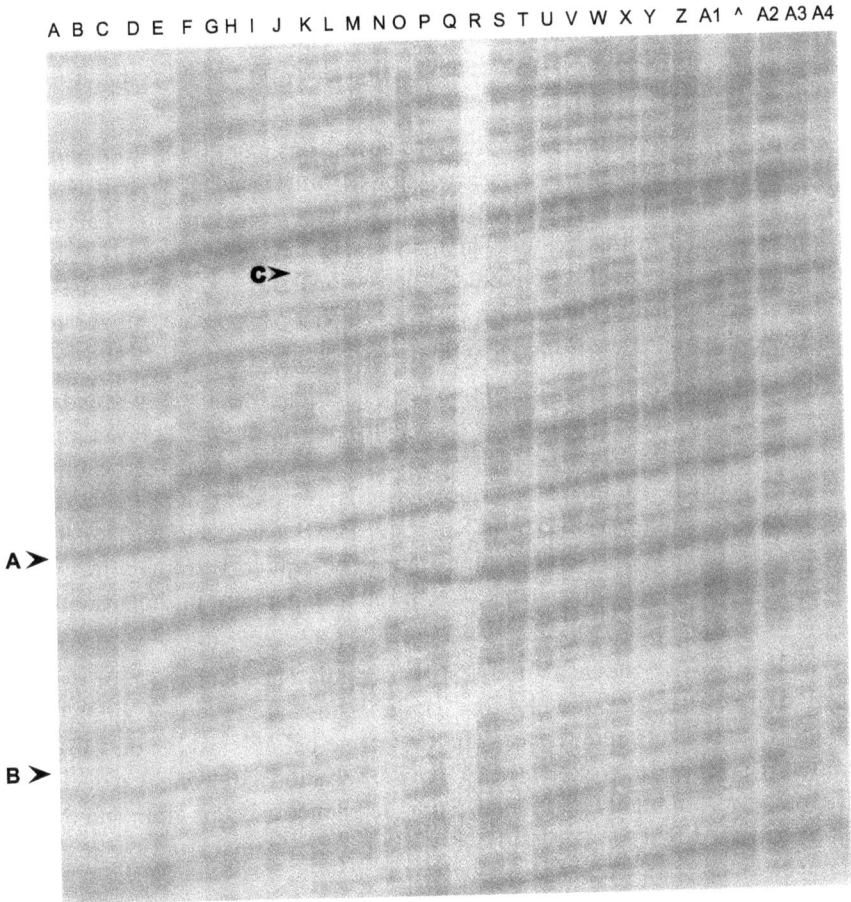

Fig. 10.4 AFLP fingerprint generated on employing primer combination E-AAG X M-CTA across the 30 amla accessions. AFLP fragments marked A is monomorphic. Bands marked B and C represent polymorphic bands that are frequent and rare, respectively.

The similarity matrix was subjected to UPGMA analysis to construct a phenetic dendogram (Fig. 10.5). The phenetic dendrogram was a near true representation of the similarity matrix as indicated by high cophenetic correlation values (r = 0.87). The overall level of genetic similarity represented by 30 amla accessions in the phenetic dendogram was 0.68. From the cluster analysis, three major groups were observed. The first major cluster comprised of accessions A, H, M and N from Bangalore, Q, R and S from Faizabad and U and V from Haryana. Within major cluster 1, two subclusters were observed. The first subcluster comprised of accessions A, H, M and N. The second subcluster, on the other hand included accessions Q, R, S, U and V.

The second major cluster comprised of accessions from Bangalore namely, B, C, D, E, F, G, I, J, L and P, two accessions from Haryana, W and X and six accessions from Delhi, Y, Z, A1, A2, A3 and A4. One accession from Faizabad (T) was also included in this group. Three subclusters were observed within major cluster 2. The first subcluster included accessions B, C, D and E. The second subcluster comprised of accessions F, G, I, X, Y, Z, A1, A2, J, W, A3 and A4. Finally, the third subcluster included three accessions namely, L, T and P.

The third major cluster included two distinct accessions, namely K and O from Bangalore that diverged out from the rest of the amla accessions at low similalarity value 0.67. The distinctiveness of these two accessions was also apparent in the fingerprint patterns. An interesting observation in the phenogram was the relative placement of the amla varieties from Faizabad. The Banarasi varieties (Q, R and S) were clustered together in the 1st major group. The Francis variety (corresponding to accession T), on the other hand, was grouped separately in the second major group.

Principal Correspondence Analysis of the amla accessions is represented in Fig. 10.6. It is clear from the plot that the accessions are relatively spread out on both the 1st and 2nd principal axes. The first principal axis explained 21% of the variation. The second axis explained 14% of the variation indicating that our original AFLP data was not highly correlated. Both the phenetic dendrogram and the plots clearly displayed that the accessions K and O were distinct from the rest of the amla accessions.

Conclusion

In summary, AFLP marker technology enabled the identification of two morphotypes of *Withania* namely Kashmiri and Nagori as genetically different. The discriminatory nature of AFLP led to the identification of a distinct genotype that was possibly a hybrid of Kashmiri and Nagori varieties although it was morphologically similar to Kashmiri. This study also revealed the presence of a considerable amount of genetic variation in *Withania*, both between and within species level. We propose that the inclusion of genotypes from a wider geographical distribution may change the genetic distance values beyond those detected in the present study, indicating an even wider genetic base.

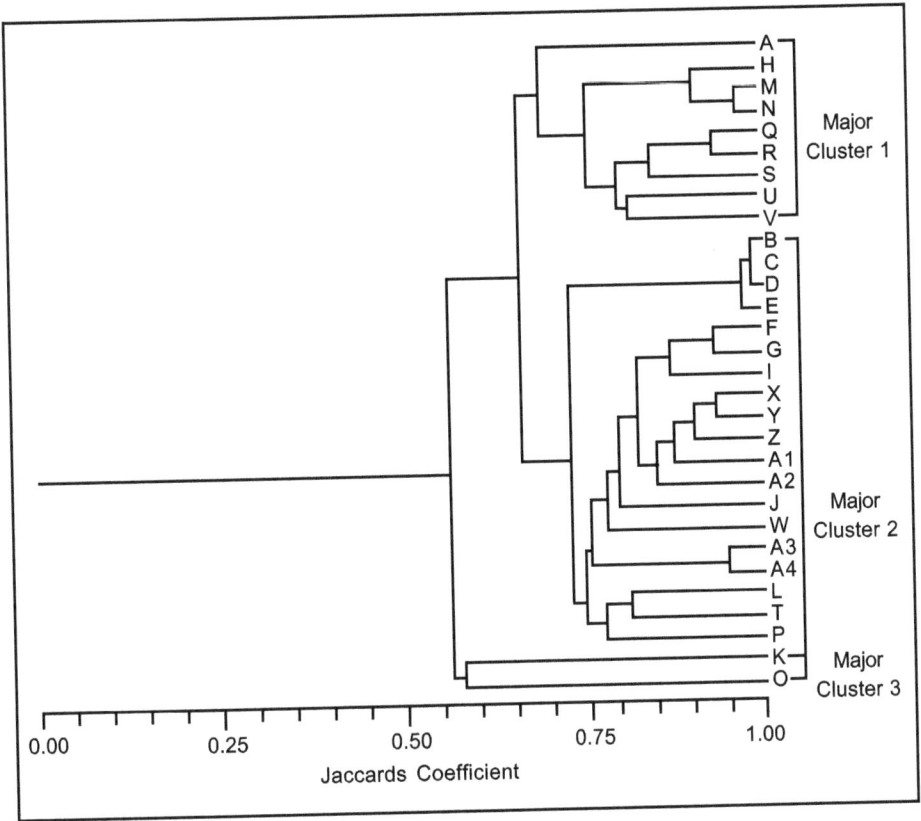

Accessions and their locations

A : Bangalore

H : Bangalore

M : Bangalore

N : Bangalore

Q : Faizabad (Banarasi)

R : Faizabad (Banarasi)

S : Faizabad (Banarasi)

U : Haryana

V : Haryana

B : Bangalore

C : Bangalore

D : Bangalore

E : Bangalore

F : Bangalore

G : Bangalore

I : Bangalore

X : Bangalore

Y : Delhi

Z : Delhi

A1 : Delhi

A2 : Delhi

J : Bangalore

W : Haryana

A3 : Delhi

A4 : Delhi

L : Bangalore

T : Faizabad (Francis)

P : Bangalore

K : Bangalore

O : Bangalore

Fig. 10.5 Phenetic dendrogram based on Jaccard's coefficient representing genetic relationships between *Phyllanthus emblica* accessions.

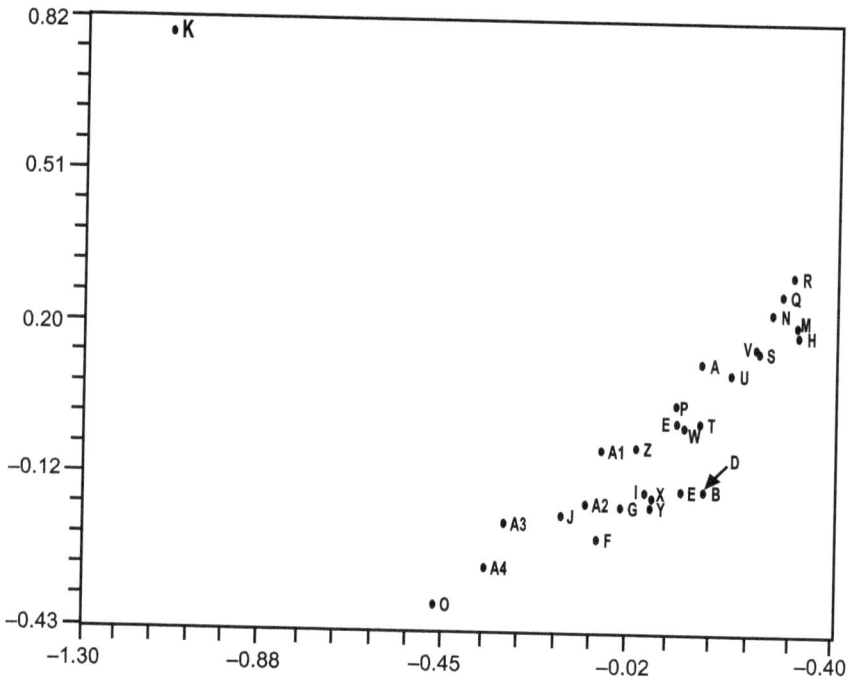

Fig. 10.6 Principal correspondence plot depicting the placement of amla accessions on a two dimensional plane based on genetic similarities.

The diversity studies in Amla germplasm revealed that there is a high level of genetic diversity in amla as reflected by low average similarity values represented in the germplasm collection. This is also indicative that this plant system has a wide genetic base. The accessions from Bangalore represented a genetically diverse stock falling in three major categories. The first group was similar to Banarasi variety from Faizabad. The second, clustered with the accessions from Haryana and Delhi. Finally, the third group was distinct from all the other amla accessions and diverged from all the other accessions at a very low similarity value. The two varieties of amla from Faizabad namely, Francis and Banarasi were identified as genetically distinct.

References

Adams MD, Kelley JM, Gocayne JD, Dubnick M, Polymeropoluos MH, Xiao H, Merril CR, Wu A, Olde B, Moreno RF, Kerlavage AR, McCombie WR and Venter JC (1991) Complimentary DNA sequencing: expressed sequence tags and human genome project. **Science** 252: 1651-1656

Aggarwal RK, Brar OS, Nandi S, Huang N and Khush GS (1999) Phylogenetic relationship among *Oryza* species revealed by AFLP markers. **Theor Appl Genet** 98: 1320-1328

Akhtar MA, Rahber BMH (1997) Antibacterial activity of plant diffusate against *Xanthomonas campestris* pv. Citri. **Int J of Pest Management** 43: 149-153

Amsellem L, Noyer JL, Le Bourgeois T and Hossaert-McKey M. (2000) Comparison of genetic diversity of the invasive weed *Rubus alceifolius* poir. (Rosaceae) in its native range and in areas of introduction, using amplified fragment length polymorphism (AFLP) markers. **Mol Ecol.** 9: 443-55

Angiolillio A, Menucuccini M and Baldoni L (1999) Olive genetic diversity assessed using amplified fragment length polymorphism. **Theor Appl Genet** 98: 411-421

Assefa S, Taliaferro CM, Anderson MP, de los Reyes BG and Edwards RM (1999) Diversity among *Cynodon* accessions based on DNA-amplification fingerprinting. **Genome** 42: 465-474

Atal CK and Schwarting (1962) Intra-specific variability in *Withania somnifera*: I. A preliminary survey. Llyodia (Cincinnatti) 25: 78-87

Bannerjee NS, Manoj P and Das MR (1999) Male-sex-associated RAPD markers in *Piper longum* L. **Curr. Sci.** 77: 693-695

Barker HAJ, Mathes M, Arnold GM, Edwards KJ, Ahmen I, Larson S and Karp A (1999) Characteristics of genetic diversity in potential biomass willows (*Salix* spp.) by RAPD and AFLP analyses. **Genome** 42: 173-183

Barlow B, Tzotsos GT (1995) Biotechnology. IN: Heywood VH, Gardner K (Eds) Global Biodiversity Assessment, Cambridge Univ Press, Cambridge, pp 671-710

Becker J, Vos P, Kuiper M, Salamini F and Heun M (1995) Combined mapping of AFLP and RFLP markers in barley. **Mol. Gen. Genet.** 249: 65-73

Beyermann B, Nurnberg P, Weihe A, Meixner M, Epplen JT and Borner T (1992) Fingerprinting plant genomes with oligonucleotide probes specific for simple repetitive DNA sequences. **Theor. Appl. Genet.** 83: **669-91**

Bhatia S, Das S, Jain A and Lakshmikumaran M (1995) DNA fingerprinting of the *B.juncea* cultivars using the microsatellite probes. **Electrophoresis** 16: 1750-1754

Bhutada SG (1999) Effect of herbal anti-stressor AV/ASE/14 and galactose payapro on milk production in buffaloes during summer. **Indian Veterinary Medical Journal** 23: 135-136

Blair MW, Panaud O and McCouch SR (1999) Inter-simple sequence repeat (ISSR) amplification for analysis of microsatellite motif frequency and fingerprinting in rice (*Oryza sativa* L.). **Theor Appl Genet** 98: 780-792

Botstein B, White R L, Skolnick M and Davis R W (1980) Construction of a genetic linkage map in man using restriction fragment length polymorphisms. **Am. J. Hum. Genet.** 32: 314-331

Broun P and Tanksley SD (1993) Characterization of tomato DNA clones with sequence similarity to human minisatellites 33.6 and 33.15. **Plant Mol. Biol.** 23: 231-242

Bukhari YM, Koivu K and Tigerstedt PMA (1999) Phylogenetic analysis of *Acacia* (Mimosaceae) as revealed from chloroplast RFLP data. **Theor. Appl. Genet.** 98: 291-298

Caranta C, Thabuis A and Palloix A (1999) Development of a CAPS marker for the *Pvr4* locus: a tool for pyramiding potyvirus resistance genes in pepper. **Genome** 42: 1111-1116

Casasoli M, Mattioni C, Cherubini M and Villani F (2001) A genetic linkage map of European chestnut (*Castanea sativa* Mill.) based on RAPD, ISSR and Isozyme markers. **Theor Appl Genet.** 102: 1190-1199

Cervera MT, Storme V, Ivens B, Gusmao J, Liu BH, Hostyn V, Van Slycken J, Van Montagu M and Boerjan W (2001) Dense Genetic Linkage Maps of Three Populus Species (*Populus deltoides, P. nigra* and *P. trichocarpa*) based on AFLP and Microsatellite Markers. **Genetics.** 158: 787-809

Charlesworth B, Sniegowski P and Stephen W (1994) The evolutionary dynamics of repetitive DNA elements in eukaryotes. **Nature** 371: 215-220

Cheung WY, Friesen L, Rakow GFW, Seguin-Swartz G and Landry BS (1997) An RFLP based linkage map of mustard [*Brassica juncea* (L) Czem and Coss] **Theor Appl Genet** 94: 841-851

Colombani VS, Causse M, Gervais Land Philouze J (2000) Efficiency of RFLP, RAPD and AFLP markers for the construction of an intra-specific map of the tomato genome. **Genome** 43: 29-40

Cresswell A, Sackville Hamilton NR, Roy A K and Viegas BM (2001) Use of amplified fragment length polymorphism markers to assess genetic diversity of *Lolium* species from Portugal. **Mol Ecol.** 10: 229-241

Dallas JF (1988) Detection of DNA fingerprints of cultivated rice by hybridization with human minisatellite DNA probe. **Proc. Natl. Acad. Sci. USA** 85: 6831-6835

Das S, Rajagopal J, Bhatia S, Srivastava PS, Lakshmikumaran M (1999) Assessment of genetic variation within *Brassica campestris* cultivars using amplified fragment length polymorphism and random amplification of polymorphic DNA markers. **J. Biosci.** 24: 433-440

Davin-Regli A, Charrel R N, Bollet C and de Mico P (1995) Variations in DNA concentrations significantly affect the reproducibility of RAPD fingerprint patterns. **Res. Microbiol.** 146: 561- 568

De Rick J, Dendauw J, Mertens M, de Loosa M, Heursel J and Van Blockstaele E (1999) Validation of criteria for the selection of AFLP markers to assess the genetic variation of a breeders collection of evergreen Azaleas. **Theor. Appl. Genet.** 99: 1155-1165

Eiadthong W, Yonemori K, Kanzaki S and Sugiura A (2000) Amplified fragment length polymorphism analysis for studying the genetic relationships among *Mangifera* species in Thailand. **J. Amer. Soc. Hort. Sci.** 125: 160-164

Elias M, Panaud O and Robert T (2000) Assessment of genetic variability in a traditional cassava (*Mannihot esculenta Crantz*) farming system using AFLP markers. **Heredity** 85: 219-30

Ellis TH, Poyser SJ, Knox MR, Vershinin A V and Ambrose AJ (1998) Polymorphism of insertion sites of Tyl-copia class retrotransposon and its use for linkage and diversity analysis in Pea. **Mol. Gen. Genomics** 260: 9-13

Erschadi S, Haberer G, Schoniger M and Torres-Ruiz RA (2000) Estimating genetic diversity of *Arabidopsis thaliana* ecotypes with amplified fragment length polymorphisms (AFLP). **Theor. Appl. Genet.** 100: 633-640

Fregne M, Bernal A, Duque M, Dixon A and Tohme J (2000) AFLP analysis of African cassava (*Mannihot esculenta* Crantz.) germplasm resistant to cassava mosaic disease (CMD). **Theor. Appl. Genet.** 100: 678-685

Garcia-Mas J, Oliver M, Panagua GH and de Vicente (2000) Comparing AFLP, RAP and RFLP markers for measuring genetic diversity in melon. **Theor. Appl. Genet.** 101: 860-864

Gaudeul M, Taberlet P and Till-Bottraud I (2000) Genetic diversity in an endangered alpine plant, *Eryngium alpinum* L. (Apiaceae), inferred from amplified fragment length polymorphism markers. **Mol. Ecol.** 9: 1625-1637

Han TH, De Jeu M, Van Eck H and Jacobsen E (2000) Genetic diversity of Chilean and Brazilian *Alstroemeria* species assessed by AFLP analysis. **Heredity** 84: 564-569

Helentjaris T, King, G, Slocum M, Siedestrang C and Wegmans S (1985) Restriction Fragment Length Polymorphism as probes for plant diversity and their tools for applied plant breeding. **Plant Mol. Biol.** 5: 109-118

Hongtrakul V, Heustis GM and Knapp SJ (1997) Amplified fragment length polymorphisms as a tool for DNA fingerprinting sunflower germplasm: genetic diversity among oilseed inbred lines. **Theor. Appl. Genet.** 95: 400-407

Jaccard P (1908) Nouvelles recherches sur la distribution florale. **Bull. Soc. Vaud. Sci. Nat.** 44: 223-270

Jeffreys AJ, Wilson V and Thein SL (1985) Individual-specific fingerprints of human DNA. **Nature** 316: 76-79

Jin H, Domier LL, Shen X and Kolb FL (2000) Combined AFLP and RFLP mapping in two oat recombinant inbred populations. **Genome** 43: 94-101

Jones CJ, Edwards KJ, Castaglione S, Winfield MO, Sala F, van de Wiel C, Bredemeijer G, Vosman B, Matthes M, Daly A, Brettscneider R, Bettini P, Buiatti M, Maestri E, Malcevschi A, Marmiroli N, Aert R, Volckaert G, Rueda J, Linacero R, Vazquez A and Karp A (1997) Reproducibility testing of RAPD, AFLP and SSR markers in plants by a network of European laboratories. **Mol.Breed.** 3: 381-390

Kiers AM, Mes TMH, Van der Meijden R and Bachmann K (2000) A search for diagnostic AFLP markers in *Chichorium* species with emphasis on endemic and chicory cultivar groups. **Genome 43**: 470-476

Konieczny A and Ausubel M (1993) A procedure for mapping *Arabidopsis* mutations using co-dominant ecotype-specific PCR-based markers. **The Plant J.** 4: 403-410

Lespinasse D, Rodier-Goud M, Grivet L, Leconte A, Legante H and Seguin M (2000) A saturated genetic linkage map of rubber tree (*Hevea* spp.) based on RFLP, AFLP, microsatellite and isozyme markers. **Theor. Appl. Genet.** 100: 127-138

Loh JP, Kiew R, Set O, Gan LH and Gan YY (2000) Amplified fragment length polymorphism fingerprinting of 16 banana cultivars (*Musa* cvs.). **Mol. Phylogenet. Evol.** 17: 360-366

Mace ES, Lester RN and Gebhardt CG (1999) AFLP analysis of genetic relationship in the tribe Datureae (Solanaceae). **Theor. Appl. Genet.** 99: 642-648

Mackill DJ, Zhang Z, Redona ED and Colowit PM (1996) Level of polymorphism and genetic mapping of AFLP markers in rice. **Genome.** 39: 969-977

Manoharan TR (1997) Biodiversity conservation in South Asia: Some regional issues. **RIS Biotechnology and Development Review** 1 : 60

McCouch SR, Chen X, Panaud O, Temnykh S, Xu Y, Cho YG, Huang N, Ishii T and Blair M (1997) Microsatellite marker development, mapping and applications in rice genetics and breeding. **Plant Mol. Biol.** 35: 89-99

Mehmood Z, Ahmad I (1999) Indian medicinal plants: A potential source for anti-candidal drugs. **Pharmaceutical Biology** 37: 237-242

Michelmore RW, Paran I and Kesseli RV (1991) Identification of markers linked to disease-resistance genes by bulked segregant analysis: A rapid method to detect markers in specific genomic regions by using segregating populations. **Proc. Natl. Acad. Sci. USA** 88:9828-9832

Morgante M and Vogel J (1994) Compound Microsatellite primers for the detection of genetic polymorphism. US Patent Appl. 08/326456

Mullis K, Faloona S, Scharf S, Saiki R, Horn G and Erlich H (1986) Specific enzymatic amplification of DNA *in vitro*. Cold Spring Harbor Symp. **Quant. Biol.** 51: 263-273

Muluvi GM, Sprent JI, Soranzo N, Provan J, Odee D, Folkard G, Mc Nicol JW and Powell W (1999) Amplified Fragment Length Polymorphism (AFLP) analysis of genetic variation in *Moringa oleifera* Lam. **Mol. Ecol.** 8: 463-470

Natesh S (2000) Biotechnology in the conservation of Medicinal and Aromatic Plants. IN: Biotechnology in Horticultural and Plantation Crops. Chandra K L, Ravindran P N and Leela Seth (eds). Malhotra Publishing House, New Delhi, India; pp 549-561

Nebauer SG, del Castillo-Agudo L and Segura J (1999) RAPD variation within and among natural populations of outcrosstng willow-leaved foxglove (*Digitalis obscura* L.). **Theor. Appl. Genet.** 98: 985-994

Negi MS, Singh A and Lakshmikumaran M (2000) Genetic variation and relationship among and within *Withania* species as revealed by AFLP markers. **Genome.** 43: 975-980

Paran I and Michelmore RW (1993) Development of reliable PCR-based markers linked to downy mildew genes in lettuce. **Theor. Appl. Genet.** 85: 985-993

Parasnis AS, Ramakrishna W, Chowdari KV, Gupta VS and Ranjekar PK (1999) Microsatellite $(GATA)_n$ reveals sex-specific differences in papaya. **Theor. Appl. Genet.** 99: 1047-1052

Peng H, Fahima T, Roder MS, Li Y C, Dahan A, Grama A, Rohin YI, Korol AB and Nevo E (1999) Microsatellite tagging of the stripe rust resistance gene *YrH52* derived from wild enmer wheat, *Triticum dicoccoides*, and suggestive negative crossover interference on chromosome 1B. **Theor. Appl. Genet.** 98: 862-872

Perrera L, Russell JR, Provan J and Powell W (2000) Use of microsatellite DNA markers to investigate the level of genetic diversity and population structure of coconut (*Cocos nucifera* L.). **Genome** 43: 15-21

Pillay M, Ogundiwin E, Nwakanma DC, Ude G and Tenkouano A (2001) Analysis of genetic diversity and relationship in East African banana germplasm. **Theor. Appl. Genet.** 102: 965-970

Prasad M, Varshney RK, Roy JK, Balyan HS and Gupta PK (2000) The use of microsatellites for detecting DNA polymorphism, genotype identification and genetic diversity in wheat. **Theor. Appl. Genet.** 100: 584-592

Rajagopal J, Bashyam L, Bhatia S, Khurana DK, Srivastava PS, Lakshmikumaran M (2000) Evaluation of Genetic Diversity in the Himalayan Poplar using RAPD markers. Silvae **Genetica** 49: 60-66

Roa AC, Chavarriaga-Aguirre P, Duque MC, Maya MM, Bonierbale MW, Iglesias C and Tohme J (2000) Cross-species amplification of cassava (*Mannihot esculenta*) (Euphorbiaceae) microsatellites: Allelic polymorphism and degree of relationship. **Am. J. Bot.** 87:1647-1655

Rodriguez JM, Berke T, Engle L, Nienhuis J (1999) Variation among and within *Capsicum* species revealed by RAPD markers. **Theor. Appl. Genet.** 99: 147-156

Rohlf F J (1995) NTSYS-pc. Numerical taxonomy and multivariate analysis system; version 1.80; Exeter publications. New York

Roldan-Ruiz I, Dendauw J, Van Bockstaele E, Depicker A and De Loose M (2000) AFLP markers reveal high polymorphic rates in ryegrasses (*Lolium* spp.). **Mol. Breed.** 6: 125-134

Russell JR, Fuller J D, Macaulay M, Hatz BG, Jahoor A, Powell W and Waugh R (1997) Direct comparison of the levels of genetic variation among barley accessions detected by RFLPs, AFLPs, SSRs and RAPDs. **Theor. Appl. Genet.** 95: 714-722

Schmidt K and Jensen K (2000) Genetic structure and AFLP variation of remnant populations in the rare plant *Pedicularis palustris* (Scrophulariaceae) and its relation to population size and reproductive components. **Am. J. Bot.** 87: 678-689.

Sharma A, Sharma R and Machii H (2000) Assessment of genetic diversity in a *Morus* germplasm collection using fluoroscent based AFLP markers. **Theor. Appl. Genet.** 101: 1049-1055

Singh A, Chaudhary C, Srivastava P S and Lakshmikumaran M (2002) Comparison of AFLP and SAMPL markers for the assessment of intra-population genetic variation in *Azadirachta indica* A. Juss. **Plant Science** 162: 17-25

Singh A, Negi MS, Rajagopal J, Bhatia S, Tomar UK, Srivastava PS and Lakshmikumaran M (1999) Assessment of genetic diversity in *Azadirachta indica* using AFLP markers. **Theor. Appl. Genet.** 99: 272-279

Sneath PHA and Sokal RR (1973) Numerical Taxonomy. W H Freeman, San Fransisco, California

Srivastava SK, Iyer SS and Ray GK (1960) Estimation of the total alkaloids of *Withania somnifera* Dunal. **Indian J. Pharma.** 22: 94

Suyama Y, Obayashi K and Hayashi I (2000) Clonal structure in a dwarf bamboo (*Sasa senanensis*) population inferred from amplified fragment length polymorphism (AFLP) fingerprints. **Mol. Ecol.** 9: 901-906

Terauchi Rand Kahl G (1999) Mapping of the *Dioscorea tokoro* genome: AFLP markers linked to sex. **Genome** 42: 752-762

Testolin R, Marrazzo T, Cipriani G, Quarta R, Verde I, Dettori M T, Pancaldi M and Sansavini S (2000) Microsatellite DNA in peach (*Prunus persica* L. Batsch) and its use in DNA fingerprinting and testing the genetic origin of cultivars. **Genome** 43: 512-520

Thomas MR and Scott NS (1994) Sequence-tagged site markers for microsatellites: simplified (technique for rapidly obtaining flanking sequences. **Plant Mol. Biol. Rep.** 12:58-64

Travis SE, Maschinski J and Keim P (1996) An analysis of genetic variation in *Astragalus cremnophylax* var. cremnophylax, a critically endangered plant, using AFLP markers. **Mol. Ecol.** 5: 735-745

Tsumura Y and Tomaru N (1999) Genetic diversity of *Cryptomeria japonica* using co-dominant DNA markers based on sequence-tagged sites. **Theor. Appl. Genet.** 98:396-404

Uma Devi P, Akagi K, Ostapenko V, Tanaka Y and Sugihara T (1996) Withaferin: A new radiosensitizer from the Indian medicinal plant *Withania somnifera*. **Indian J. Radiat. Bioi.** 69: 193-197

Uma Devi P, Sharda A C and Emerson Solomon F (1993) Anti-tumour and radiosensitizing effects of *Withania somnifera* (Ashwagandha) on transplantable mouse tumour, Sarcoma-180. **Indian J. Exp. Biol**. 31: 607-611.

Uma Shaanker R and Ganeshaiah K N (1997) Mapping genetic diversity of *Phyllanthus emblica*: Forest gene banks as a new approach for *in situ* conservation of genetic resources. **Curr. Sci.** 73: 163-168

Van Der Hulst RG, Mes TH, Den Nijs JC and Bachmann K (2000) Amplified fragment length polymorphism (AFLP) markers reveal that population structure of triploid dandelions (*Taraxacum officinale*) exhibits both clonality and recombination. **Mol. Ecol.** 9:1-8

Van Der Stapper J, Weltjens I, Van Campenhout S and Volckaert G (1999) Genetic relationship among *Stylosanthes* species as revealed by sequence-tagged site markers, **Theor. Appl. Genet.** 98: 1045-1062

Vogel, JM and Scolnick, PA (1998) Direct amplification from microsatellite- Detection of simple repeat based polymorphism without cloning. IN: Cateno-Anolies, G and Gresshoff, P.M (Eds) DNA markers, Protocols and Applications. Wiley-Liss Inc. New York-pp. 133-150.

Voorrips RE, Jongerius MC and Kanne HJ (1997) Mapping of two genes for resistance to clubfoot (*Plasmodiophora brassicae*) in a population of doubled haploid lines of *Brassica oleracea* by means of RFLP and AFLP markers. **Theor. Appl. Genet.** 94: 75-82

Vos P, Hogers R, Bleeker M, Reijans M, van de Lee T, Hornes M, Frijters A, Pot J, Peleman J, Kuipers M and Zabeau M (1995) AFLP: A new technique for DNA fingerprinting. **Nucl. Acids Res.** 4407-4414

Weising K, Weigand F, Driesel A, Kahl G, Zischler Hand Epplen J T (1989) Polymorphic simple GATA / GACA repeats in plant genomes. **Nucl. Acids Res.** 17: 10128

Welsh J and McClelland M (1990) Fingerprinting genomes using PCR with arbitrary primer. **Nucl. Acids Res.** 18: 7213-7218

Welsh J and McClelland M (1991) Genomic fingerprinting using arbitrarily primed PCR and a matrix of pair-wise comparison of primers. **Nucl. Acids Res.** 19: 5277-5279

Westman, AI and Kresowich, S (1997) Use of molecular marker techniques for description of plant genetic variations. IN : Callow J. A, Ford-Lloyd, BV and Newbury H.J (Eds) Biotechnology and Plant Genetic Resources-Conservation and use. CAB International, Reading U.K. pp 9-48.

Williams GK, Kubelik AR, Livak K J, Rafalski J A and Tingey S V (1990) DNA polymorphism amplified by arbitrary primers are useful as genetic markers. **Nucl. Acids Res.** 18: 6531- 6535

Zabeau M and Vos P (1993) Selective restriction fragment amplification: A general method for DNA fingerprinting. European patent Application of 92402629 (Publ. No. 0534858A 1)

Zhang YJ, Tanaka T (2000) Phyllaemblic acid, a novel highly oxygenated norbisabolane from the roots of *Phyllanthus emblica*. **Tetrahedron Letters** 41: 1781-1784

Zheng K, Qian H, Shen B, Zhuang J, Un Hand Lu J (1994) RFLP-based phylogenetic analysis of wide compatibility varieties in *Oryza saliva* L. **Theor. Appl. Genet.** 88: 65-69

11

Marker Assisted Selection for Improving Abiotic Stress Tolerance in Dryland Crops

N. Seetharama
National Research Centre for Sorghum, Rajendranagar, Hyderabad-500 030

Drought, soil salinity and acidity are the three major factors that limit plant growth and productivity by about one-third of the potential economic yields of crops. For the purpose of this discussion we will ignore other abiotic stresses such as temperature and mineral stresses (both the effect of excess and deficits), which are less complex to understand, and differ somewhat in terms of practical steps to ameliorate their effects on crop production. (Nevertheless, one is always conscious of interactions among different types of abiotic stresses and their further interactions with biotic stresses in crop production, especially under dryland conditions). It is especially true in a country like India where the scope for enlarging cultivable land under irrigation is limited, and where the livelihood of a large section of the population is wholly dependent on dryland agriculture.

In spite of the surge in literature on drought resistance in crops during the past two decades, practical progress in breeding for drought tolerance is not appreciable. While there can be different approaches, considering the fact that drought traits are many, and they highly interact with environment, the DNA marker based selection methods deserve much attention. It requires an analytical approach of dissecting and studying the contribution of component traits using the quantitative trait loci (QTL, generally used both as singular and plural, strictly only plural) model. The underlying assumptions are: (i) a sufficiently dense map is available, (ii) the evaluation of the phenotype (relative agronomic performance under drought, measurement of specific morphological or physiological trait, expression of specific response or metabolite, etc) is precise, and (iii) genotype x environment (eventually trait, or even marker x environment) interactions can be quantified. Considering the complexity of drought and often the conflicting requirements for traits of the same species to be grown under different environments, it is important to phenotype under well-managed and quantified nursery or greenhouse conditions, along with large-scale field evaluations under natural drought in the target region. A strong demonstrated linkage between traits in relation to crop performance under the target environment and the DNA marker is a crucial step before advocating marker-assisted selection (MAS). Once robust markers are found, they can be used for screening germplasm, selection of the desired traits in breeding, and gene pyramiding.

Maps and Markers

Genetic linkage maps of over 30 species are already developed, and more are under development. Most of the maps are developed for annual crops, but maps of perennials like pine are also under rapid progress. Among them, the maps of only few crops {apart from that of model species (*Arabidopsis*), that of rice (model crop) and maize (genetics well-studied)} are saturated with sufficient number of markers (2000 or more). Some are being annotated rapidly. However, in the light of knowledge of DNA sequence of whole genome in rice, rapid progress can occur, especially in the context of syntenic relationships in crops. Again, only in few cases the correspondence between linkage and physical maps are established. Similarly, the location of telomeres and centromers, or the relationship between individual chromosome (named by cytzogeneticist) and linkage groups is known only for important crops like maize and rice. Ease of developing and characterizing BAC libraries, rapid automated sequencing and annotating tools, progress in techniques like *in situ* hybridization, comparative mapping - will all allow other important crops like sorghum to catch up with the model crops.

Application of molecular genetic diagnostics in plant breeding is still in its infancy. This is particularly so with regard to selection for abiotic stress tolerance. For efficient and accurate identification of chromosomal region with major phenotypic effect (QTL peak) densely saturated maps and large mapping population are required. Major progress has been made in identification of important crops like maize, barley, rice, potato, wheat and tomato. In barley highly significant QTL have been found for agronomic traits like yield and lodging, and these have been found on different linkage groups, somewhat distributed evenly. Some are stable across environments, and most of the inconsistency in environment x QTL interactions is largely related to the size of QTL effects. Careful consideration of the following is critical for the success of any marker-assisted selection program:

1. Choice of parents in terms of differences between them for both the target trait and in the genome (DNA), mapping populations types and size.
2. Marker types, cost-effectiveness, data analysis (software)
3. Ability to phenotype accurately over a set of target environments or managed nurseries
4. Number of QTLs involved, the ease of pyramiding them and testing the effect of each.

Summary of Published Literature on Identification of Markers and their Use

While QTL analysis of drought-resistance traits is encouraging, the challenge of breeding for drought resistance using DNA markers is still at its infancy. Progress will come in incremental steps, but atleast in the near future it will be confined to specific crops under well-defined stress patterns. Detailed review is beyond the scope of this paper. However, few recent conference proceedings and some of the highlights extracted from the recent literature are here with provided.

1. *Traits chosen for finding markers, and the relationships between traits themselves :* As expected, most studies dealt with specific components that are likely to be significant for crop growth and yield under drought. They include morphological (e.g., root and shoot characteristics in rice), developmental (anthesis to silking interval, ASI, in maize), physiological (osmotic adjustment, OA), metabolic (ABA accumulation), and agronomic (water-use efficiency, WUE) characteristics. Several studies also included yield or yield components. Identifying QTL for both physiological components of drought resistance traits and final yield and yield components under drought in the same study is rare.

2. The number of QTLs detected for a trait generally varied between 1 and 4 or even more, and QTLs were spread across the genome on several linkage groups (LG, or chromosomes, ch.). In a few cases, there were useful QTLs for more than one trait on the same LG (e.g., osmotic adjustment and dehydration tolerance in the region associated with root morphology in rice.

3. Phenotypic variation for the measured trait accounted for by an individual QTL was generally about 10% however some exceptions are noted. For example, the QTL for root length at 28 days after sowing in rice accounted for 30% of phenotypic variation, but such observations can only be taken seriously if they are repeatable. The best model (with about 3 QTLs) rarely accounted for more than 50% of the observed phenotypic effect.

4. In several cases, association between QTLs for different traits existed. Many of them may represent interrelationships between traits themselves. This is especially true for yield components. Pleiotropic effects were also common.

5. Measuring integrated responses by means of traits such as water-use efficiency indicates involvement of only a few QTLs (4 - 5 in soybean) in different populations. QTLs for carbon discrimination (indicating water-use efficiency) were found in both wheat and tomato.

6. Often both parents (including the drought - "susceptible" one) contribute QTLs for the measured response. Thus, detection of transgressive segregants is common in many mapping populations.

7. Localization and co-localization of QTLs for different measured traits can be easily interpreted, such as those for relative water content and number of leaves per tiller in barley, even when not all QTLs for each trait are located in the same region.

Stability of QTL and Relationship between Yield Under Stress and QTL

Stability of QTL across enviromnents or stress levels is rarely studied except for yield or yield components. In the case of maize, in spite of >10-fold differences in yields, the QTL for yield were consistent across enviromnents. Evidence of crossover-type interactions across generations tested under different stress levels was only slight, and most of the interactions were in the form of a change in magnitude of QTL effects. Only a few studies deal with identifying QTLs across different mapping populations. Not all QTLs were shared between them even when one parent was common.

In some cases, the relationship between QTL and yield was negative, but in others, such as ASI (maize), the drought-resistance trait did not decrease yield under well-watered conditions. QTLs such as ASI, which are stable over years and stress levels are the immediate candidates for use in MAS in practical breeding are already used extensively in maize. As the experimental precision decreased under severe (water or nitrogen) stress, fewer QTLs were detected, which is comparable with the observations under low nitrogen as has been observed by breeders for long. For correct interpretation of QTL data, measurement of phenotypic values in relation to specific and measured characteristics at test-environments is as important as genotypic data.

Other Considerations for Detecting QTLs and Assessing their Effects, Especially on Yield

Many of the QTLs associated with drought-resistance traits may or may not be associated with yield potential. Further, there may be major inconsistencies in genomic positions in QTL across water regimes as noted for maize. In maize, the yields of test materials (RI lines backcrossed to each parent) were tenfold less under stress than control, but the QTL for yield were nearly the same. Breeders could rely on mapping data from favorable environments for breeding for materials adapted to stress environments. Only few researchers find little evidence of crossover-type interactions across generations and stress levels. Instead, the interactions seemed to be in the form of a change in magnitude of effects, although only 17% of the QTL were common across stress environments. This observation is promising for the use of QTL. Further, common QTL across generations and stress levels could be selected during early generations, whereas in later generations only field evaluations may be used to select for specific adaptability.

Additional Approaches

An alternate and popular approach is to create a novel and functionally known type of variability in plant stress response by genetic transformation. The transgenic approach offers a powerful means of incorporating a broad spectrum of genes with profound ability to up - or down-regulate specific metabolic steps associated with stress response. But transformation with any single gene (including transcription factors) or group of genes for a particular pathway may not be adequate for conferring drought resistance because the products from several interacting pathways (web) are required to ensure drought resistance. The transgenic approach is useful, however, to obtain valuable information on the test crop and hypothesis in relation to specific steps in alleviation of drought stress. This approach will thus be useful in identifying candidate genes for stress resistance or its components with significant developments.

The revolutionary stage of functional genomics is yet to begin and is awaiting the dust of excitement to settle. In the immediate future, the drought researchers can benefit from the identification of suitable markers for drought tolerance. At the moment, this need is partially satisfied by comparative mapping. So far the classical (molecular biological) techniques such as identification of candidate genes using differential libraries and displays (DDRT) have been of limited use. The chip technology can be far more effective, especially if one is targeting specific and easily quantifiable traits only.

Drought Physiology and QTL Approach

For drought resistance, one is dealing with large number of traits and QTLs. Therefore, there is scope to look for many clues regarding adaptive mechanisms, useful genes, environmental effects, and further scope for finding more markers. Physiologists need to review the causal relationship between the co-localized QTLs and the associated traits using the tools of molecular genetics. If reasonable assumptions on the relationships among the component traits can be made, interpretation may not be difficult. Since QTLs for many traits can be identified from the mapping population, and many such QTLs can be found in close vicinity on the same chromosome, targeting few traits simultaneously are beneficial.

If a QTL maps to a region close to genes of known function, the physiologist's goal of designing a drought-tolerant plant and the breeder's goal of finding the most-effective DNA markers are met to a significant extent.

Finally, considering the complexity of drought, and the costs involved, there is ample scope for greater participatory approach for breeding, especially for meeting local needs. For example, the marker-assisted selection for rust resistance in popular varieties of rabi sorghum coupled with on-farm trials using farmer participatory approach can identify useful genotypes for immediate use in dryland crop production. In the end, considering the challenges of dryland agriculture, we need to be very pragmatic and cost-conscious in our approach, even if the priests of high science may label them as poor compromises.

12

Plants as Bioreactors for Producing Vaccines

K.V. Rao and V. D. Reddy
Centre for Plant Molecular Biology Osmania University, Hyderabad-500 007

Recent advances in molecular biology and immunology have led to development of effective vaccines and promising strategies for production of new recombinant vaccines for infectious human diseases. The traditional vaccines now in use are based on the intact disease-inducing agents-inactivated or live attenuated or inactivated toxins. In many cases, these approaches have been very useful at inducing immune protection, mainly based on antibody responses and have led to diminished incidence in morbidity and mortality of a large number of infectious diseases. Yet there are several drawbacks incurred by the current procedures for vaccine preparation, such as, difficulty of *in vitro* culturing of several viruses and parasites, biohazard and safety considerations, and components that can cause undesirable effects as well as the loss of efficacy due to genetic variation of many pathogens (O'Hagan *et al.*, 2001). As a result of these problems, several new approaches for vaccine development have emerged, which may have significant advantages over traditional approaches which include: (i) recombinant protein subunits; (ii) synthetic peptides; (iii) protein polysaccharide conjugates; and (iv) plasmid DNA (PNA vaccines).

Vaccination is a cost-effective tool for disease prevention. Vaccines have been developed for many infectious diseases including major devastating diseases such as, diphtheria, pertussis, (whooping cough), polio, measles, tetanus and tuberculosis. Millions of infants continue to succumb to these infectious diseases because they are unreachable to most remote and undeveloped parts of the globe (Langridge, 2000). There is a need to improve the safety and efficiency of vaccines currently in use and to find ways to lower their cost of production and deliver them efficiently.

The rapid advances in recombinant DNA technology, basic molecular biology and the development of genetic transformation techniques for plants have increased the value of plants with a large number of applications in agriculture, industry and medicine. One of the fast growing applications of biotechnology is the use of crop plants as bioreactors for the commercial scale production of recombinant proteins (Giddings *et al.*, 2000; Daniell *et al.*, 2001). Transgenic plants expressing antigens induce immune responses and have several benefits over current vaccine technologies, including increased safety, economy, stability, versatility and efficacy (Streatfield *et al.*, 2001). Apart from economic advantages, there are qualitative benefits favouring the use of transgenic plants as bioreactors for producing these recombinant proteins. The products from transgenic plants are unlikely to be contaminated by animal pathogens, microbial toxins or oncogenic sequences (Fisher and Emans, 2001). Introduction and expression of genes encoding antigenic epitopes for viral, bacterial and other pathogens in plant system would be advantageous for making vaccines available at an affordable cost for the population of developing countries.

Plants as Bioreactors for Producing Vaccines

Increasing knowledge of plant molecular biology and vaccines has opened up possibilies of using plant as a system for producing edible vaccines. Many infectious agents colonize or invade epithelial membranes and are transmitted through the contaminated food or water. Vaccines that are effective against these infections must stimulate the mucosal immune system. In general, a mucosal immune response is more effectively achieved by oral rather than parenteral antigen delivery route. Several particulate antigens have proven to be effective oral immunogens including live and killed microorganisms. Subunit vaccines based upon recombinant cell-culture expression systems are feasible but, for commercial scale production, these systems require fermentation technology and stringent purification protocols so that sufficient quantities of recombinant protein can be obtained for oral delivery. Transgenic plants that express antigens in their edible tissue can be used as an inexpensive oral-vaccine production and delivery system; therefore, immunization might be possible simply through consumption of an edible vaccine. Plants are modified on the basis of peptide epitopes of pathogens to produce vaccines against a variety of human and animal diseases. Stable transformation is characterized by the integration of the target gene into the plant genome. Once the gene is transformed successfully into the plant for an immunogenic subunit, the plant system produces the vaccine. Oral delivery of vaccines is an attractive alternative to injection, largely for reasons of low cost and easy administration. These oral vaccines will induce both mucosal and systemic immune responses as desired (Arakawa *et al.*, 1998).

The major breakthrough for using plants as bioreactors for producing vaccines occurred when researchers established that plants could indeed synthesize foreign antigen in their proper confirmation. The first report on the production of vaccine by expressing a surface protein from *Streptococcus* in tobacco at a concentration of 0.02% of the total leaf protein was published in 1990 in the form of patent application under the International Patent Cooperation Treaty. In the early 1990 Charles Arntzen conceived the idea that food could be genetically engineered to produce vaccines, which can be consumed when, inoculations were needed. Initially he expressed hepatitis B virus antigenic genes in tobacco mainly to see whether the expressed recombinant proteins have the ability to induce immune system in mice (Haq et al., 1995). Despite some difficulties in the utilization of plants, first human clinical trials are promising. By developing an edible vaccine, researchers focus on pathogens, which invade the host via the mucosal surfaces, such as the gastrointestinal and respiratory tracts. Candidates include the Norwalk and Rotavirus, which cause severe diarrhoea.

The second choice is the right plant for the job. The plant material should be eaten raw to avoid degradation of its antigen by cooking. Banana would be an ideal fruit, loved by young and old, and harboring a fairly high protein content. However, the reproducible transformation system for banana is still not available and takes almost two years from transformation to fruit development. Alternatively the two common vegetables namely potato and tomato are ideal plant systems because they could be transformed routinely and grow faster, for development of edible vaccines.

To evolve a plant type which can produce a viral or bacterial antigen, the genetic information must be brought into a readable context. Today many different plant promoters and regulatory sequences are known enabling targeted or inducible expression of the foreign gene. For the introduction of plasmid-containing antigen of interest into the plant, usually *Agrobacterium*-mediated transformation or biolistic delivery with the gene gun is applied. In both the cases, the DNA is integrated into the plant genome on a random basis, which results in a different expression level for each independent line. Screening for the best-expressed line is necessary for optimizing antigen production. Consumable common foods are on trial, as alternatives to injectable vaccines including banana, potato and tomato as well as lettuce, rice, wheat, corn, soya bean etc. (Daniell, et al., 2001). Consumption of fruit as processed food would provide an immune response and vaccination.

Plants obtained after transformation are, in general, hemizygous for an inserted gene. To obtain homozygous lines the plants need to be self-pollinated. By crossing two plant lines harboring different antigens, multi component vaccines could be obtained. This feature is another advantage of plants over other vaccine production systems.

A large number of antigens have been successfully expressed in plants (Table 12.1). Recent studies have shown that the antigens of pathogens causing severe diarrhoea have demonstrated their potential as plant expressed oral vaccines: the LT-B subunit of enterotoxigenic *E. coli* (Tacket et al., 1998;

Table 12.1 Production of recombinant vaccines in transgenic plants

Protein source	Protein expressed	Plant expression system	Protective capacity of the vaccine	References
Enterotoxigenic E. coli (humans)	Heat-labile toxin B-sub unit	Tobacco, Potato and Maize	Immunogenic-orally	Haq T.A. et al., 1995. Mason, H.S. et al., 1998. Tacket, C.O. et al., 1998.
Vibrio cholera (humans)	Cholera toxin B-sub unit	Potato	Immunogenic-orally	Arakawa, T. et al., 1997. Arakawa, T. et al., 1998.
Hepatitis B virus (humans)	Envelope surface protein	Tobacco, Potato, Carrot, Lupin and Banana	Immunogenic-injection.	Mason et al., 1992. Thanavala, Y., et al., 1995. Richter, L.J. et al., 2001. Kapusta, J. et al., 1999.
Measles virus	Hemagglutinin protein	Tobacco	Immunogenic-orally.	Huang, Z. et al., 2001.
Norwalk virus (humans)	Capsid protein	Tobacco and Potato	Immunogenic-orally.	Mason, H.S. et al., 1996. Tacket, C.O. et al., 2000.
Respiratory syncytial virus	Syncytial virus F-protein	Tomato	Immunogenic-orally	Sandhu, J.S. et al., 2000
Rabies virus (humans)	Glycoprotein	Tomato	Immunogenic-orally	McGarvey, P.B. et al., 1995. Modelska, A. et al., 1998.
Human cytomegalovirus (humans)	Glycoprotein B	Tobacco	Immunologically related protein	Tackaberry, E.S. et al., 1999.
Rabbit hemorrhagic disease virus (rabbits)	VP60	Potato	Immunogenic-orally	Castanon, S. et al., 1999.
Foot & mouth disease virus of Livestock	Epitope-VP1	Arabidopsis and Alfalfa	Immunogenic-injection.	Carrillo, C. et al., 1998. Wigdorovitz, A. et al., 1999.
Transmisable gastro enteritis coronavirus (pigs & swine)	Glycoprotein (S)	Arabidopsis, Tobacco and Maize	Immunogenic-injection	Gomez, N. et al., 1998. Tuboly, T et al., 2000. Streatfield, S.J. et al., 2000.

Table 12.1 Contd . . .

Protein source	Protein expressed	Plant expression system	Protective capacity of the vaccine	References
Autoimmune diabetes	Cholera toxin B-human insulin	Potato	Immunogenic-orally	Arakawa, T. *et al.*, 1998.
Dental caries	Streptococcus surface protein (SpaA)	Tobacco	Immunogenic	Tacket, C.O.. and Mason, H.S., 1999.
Malaria	Malarial B cell epitope	Tobacco		Turpen, T.H., *et al.*, 1995. Tacket, C.O. and Mason, H.S., 1999.
HIV	HIV epitopes Gp 120 and 41	Tobacco/ Cow-pea		Brennan, F.R. *et al.*, 1999. Durrani, Z. *et al.*, 1998.
Cancer	c-Myc	Tobacco		Beachy, R.N. et al., 1996.

Lauterslager *et al.*, 2001) and the Norwalk virus capsid protein (Mason *et al.*, 1996; Tacket *et al.*, 2000). Both have passed phase II trails. Also, the hepatitis B surface antigen (HBsAg), and measles virus hemagglutinin protein, proved that the edible plant tissue is sufficient to protect the antigen against digestion (Kapusta *et al.*, 1999; Richter *et al.*, 2001; Huang *et al.*, 2001). Respiratory syncytial virus F protein (Sandhu *et al.*, 2000), and the rabies virus G-protein (McGarvey *et al.*, 1995; Modelska *et al.*, 1998) antigens have also been expressed. These antigens have induced systemic and mucosal immune responses without aid of adjuvants and no adverse effects of genetically modified materials have been demonstrated. Conjugation of these therapeutic molecules with CT-B (Cholera toxin-B) or LT-B (*E.coli* heat labile enterotoxin) will greatly facilitate antigen delivery and presentation to the gut associated lymphoid tissue (GALT) due to its affinity for the cell surface receptor GM_1-gangloside located on cells of the GALT (Arakawa, *et al.*, 1998). This would result in increased uptake and immunologic recognition.

One of the major limitations of the expression of recombinant antigens in transgenic plants remains the achievement of a high yield that is sufficient to confer total protection in humans. A dramatic increase in the recombinant protein yields in plants can be achieved by chloroplast transformation. Using this approach, recombinant proteins can be expressed with yields reaching more than 8-10% of total soluble protein. Another major concern about glycoprotein expression in plants is the presence of plant specific glycons that might alter the properties of the recombinant protein. Strategies to humanize plant N-glycons have been developed recently, including inhibition of endogenous golgi-glycosyl-transferases or addition of mammalian glycosyl-transferases.

Progress Achieved at Centre for Plant Molecular Biology (CPMB)

Synthesis and expression of a gene coding for different epitopes of surface antigen of Hepatitis B Virus

Infection with hepatitis B virus (HBV) causes serious liver diseases in humans. Hepatitis B surface antigen (HBsAg) of HBV is being used widely in the immunization programs. The subunit vaccines developed so far, are not effective against all the four strains (*adw/adr/ayr/ayw*) of hepatitis B virus. Hepatitis B virus surface antigens have been expressed in bacterial, yeast and mammalian expression systems. A novel gene was designed and synthesized (HBsAg), comprising three antigenic epitopes, DPRVRG representing preS2 region of *ayr, ayw* and *adr* strains, STGPCK and KPTDGN representing S region of adw viral strain. The synthetic HBsAg gene was subcloned into bacterial expression vector pET 2Ib(+) and expressed *in vivo* and *in vitro*. Synthetic gene encoded HBsAg protein when injected into Balb/c mice elicited an immune response. The synthetic HBsAg gene encoded polypeptide reacted with the antisera raised against natural HBsAg antigen suggesting its suitability as an antigen. Immunogenicity and antigenicity exhibited by the synthetic gene encoded polypeptide amply indicates that it can be used as an effective recombinant subunit vaccine for controlling infection caused by the hepatitis B virus. Also the synthetic HBsAg gene was subcloned into pCI-neo mammalian expression vector and Balb/c mice were immunized using this plasmid. Antisera obtained against DNA-based immunization reacted with synthetic gene encoded protein expressed in *E. coli* suggesting that mammalian expression vector containing synthetic gene has great potential in developing an effective recombinant vaccine against HBV strains.

To elevate the immunogenic response of the synthetic gene, the HBsAg gene is fused with the CT-B (Cholera Toxin-B) sub unit isolated from *Vibrio cholera*. The genomic DNA isolated from *Vibrio cholera* was used to isolate the CT-B gene by Polymerase chain reaction (PCR) using the CT-B gene specific primers. The amplified DNA fragment (374 bp) was sequenced and confirmed the coding sequence of CT-B gene. The isolated gene, CT-B, was cloned into the bacterial expression vector pET2Ib(+) containing the HBsAg gene to express as fusion protein. The expression of fusion protein was confirmed in *E. coli*.

Agrobacterium mediated genetic transformation studies in tomato

Tissue culture conditions and genetic transformation protocol for the tomato varieties namely, NTV61 I, NTV612, NTVI06, NTV123 and NTV32 were established mainly to introduce and express the novel synthetic gene (HBsAg) constructed for hepatitis B virus. Callus was initiated from first leaf of four week old plants on MS medium supplemented with 3.0% sucrose and various combinations of BAP, Kinetin, NAA and IAA. High frequency of callus induction was observed on MS medium containing 3.0% sucrose, BAP/Kinetin 1.0mg/l, and NAA/IAA 05-2.0mg/l.

115

In order to standardize the *Agrobaclerium* mediated transformation system for tomato, one month old leaf derived calli were used for transformation studies. *Agrobacterium* vectors, pTOK233 and pKIWI I OS, containing gusA reporter gene were employed. After transformation the calli was subjected to selection medium containing kanamycin or hygromycin (30-50mg/l). The actively growing calli were selected and transferred to proliferation medium. The embryogenic portions of the calli were selected and transferred to regeneration medium. The regenerated plants were transplanted in pots and allowed to grow to maturity.

Future Prospects

The number of medically relevant molecules produced in transgenic plants is increasing exponentially, from recombinant antibodies to oral vaccines. There are still improvements to be made in areas such as level of expression and glycosylation. The potentially low cost of production and scale up to agricultural levels that plants promise, should provide a source for antibodies, vaccines and therapeutic molecules for the population of the whole world. With these advances, it seems likely that the molecular farming becomes an economic and agricultural reality in the near future.

References

Arakawa, T, Chang, DKX and Langridge, WHR. (1998). Efficacy of a food plant-based oral cholera toxin B subunit vaccine. **Nature Biotech.** 16: 292-297.

Arakawa, T, Chang, DKX., Merritt, J.L. and Langridge, WHR. (1998). Expression of cholera toxin B subunit oligomers in transgenic potato plants. **Transgenic Res.** 6,403-413.

Arakawa,T, Yu., J,. Chong, D.K, Hough, J, Engen, P.C and Langridge, W.H.R. (1998). A Plant-Based Cholera Toxin B Subunit-Insulin Fusion Protein Protects against the Development of Autoimmune diabetes. **Nature Biotech.** 16(10): 934-938.

Arntzen, C.J. (1997) High-tech herbal medicine: plant-based vaccines. **Nature Biotech.** 15: 221-222.

Artzen, C. J. and Mason, H.S. (1997) oral vaccine production in the edible tissues of transgenic plants. IN. Levin, M.M. Woodrow, G.C. Kaper, J.B and Cobon, G.S. (Eds) New Generation Vaccines. Marcel Dekker Inc. Newyork pp 263-277.

Beachy, R. N. Fitchen, J. H. and Hein, M. B. (1996). Use of plant viruses for delivery of vaccine epitopes. In: Engineering plants for commercial products and applications, 43-49.

Brennan, F.R., Bellaby, T., Helliwell, S.M., Jones, T.D., Kamstrup, S., Dalsgaard, K., Flock, J.I and Hamilton, W.D.O. (1999). Chimeric plant virus particles administered nasally or orally induce systemic and mucosal immuno responses in mice. **J. Virol.** 73: 930-938.

Chargelegne, D, Obregon, P and Drake, PMW (2001). Transgenic plants for vaccine production: Expectations and limitations. **Trends in Plant Sci.** 6: 495-496.

Dalsgaard, K.., Uttenthal, A., Jones, T.D., Xu, F., Merryweather, A., Hamilton, WDO, Langeveld, J.P.M., Boshuizen, R.S. Kamstrup, S., Lomonossoff, G.P., Porta, C., Vela, C., Casal, J.I., Meloen, R.H. and Rodgers, P .B. (1997). Plant-derived vaccine protects target animals against a viral disease. **Nature Biotech.** 15: 248-252.

Daniell, H., Streatfield, S.J. and Wycoff, K. (2001). Medical molecular farming: Production of antibodies, biopharmaceuticals and edible vaccines in plants. **Trends in Plant Sci.** 6: 219-226.

Durrani, Z. (1998). Intranasal immunization with a plant virus expressing a peptide from HIV-I gp41 stimulates better mucosal and systemic HIV-I-specific IgA and IgG than oral immunization. **J. Immunol. Methods.** 220: 93-103.

Fisher, R and Emans, N. (2000). Molecular farming of pharmaceutical proteins. **Transgenic Res.** 9: 279-299.

Giddings, G., Allison, G., Brroks, D. and Carter, A. (2000). Transgenic plants as factories for biopharmaceuticals. **Nature Biotech.** 18(11): 1130.

Haq, T.A., Mason, H.S., Clements, J.D., and Arntzen, C.J. (1995). Oral immunization with a recombinant bacterial antigen producing transgenic plants. **Science.** 268: 714-716

Huang, Z, Dry, I. and Webster, D (2001). Plant derived measels virus hemagglutinin protein induces neutralizing antibodies in mice. **Vaccine,** 19: 2163-2171.

Kapusta, J., Modelska, A., Figleronicz, M., Pniewski, T., Letellier, M., Lisowa, O., Yusiba, V., Kaprowski, H., Plucienniczak, A., Legocki, A.B.(1999). A plant derived edible vaccine against hepatitis B virus. F ASEB, 13, 1996-1999.

Langridge, W.H.R. (2000). Edible vaccines. **Scientific American.** 283: 48-53.

Lauterslager, T.G.M., Lauterslager, D.E.A., Florack, van der Wal, T.J., Molthoff, J.W., Langeveld, J.P.M., Bosch, D., Boersma, W.J.A. and Hilgers, L.A.T. (2001). Oral immunization of naive and primed animals with potato tubers expressing LTB. **Vaccine** 19: 2749-2755.

Lerouge, P. (2000). N-glycosylation of recombinant pharmaceutical glycoproteins produced in transgenic plants: Towards a humanization of plant N-glycons. **Curr. Pharmacol. Biotechnol.** 1: 347-354.

Mason, H.S., Ball, J.M., Shi, J.J., Jiang, X., Estes, M.K. and Arntzen, C.J. (1996). Expression of Norwalk Virus Capsid Protein in Transgenic Tobacco and Potato and its Oral Immunogenicity in Mice. **Proc. Natl. Acad. Sci. USA,** 93: 5335-5340.

Mason, H.S., Lam, D.M.K. and Arntzen, C.J. (1992). Expression of hepatitis B surface antigen in transgenic plants. **Proc. Natl. Acad. Sci. USA,** 89: 11745-11749.

McGarvey, P.B., Hammond, J., Dienelt, M.M., Hopper, D.C., Fu, Z.F., Dietzschold, B., Koprowski, H. and Michaels, F.H. (1995). Expression of the rabies virus glycoprotein in transgenic tomatoes. **Bio/Technology.** 13, 1484-1487.

Modelska, A., Dietzschold, B., and Sleysh, N. (1998). Immunisation against rabies with plant derived antigen. **Proc. Natl. Acad. Sci.** USA 95: 2481-2485.

O'Hagan, D.T., MacKichan, M.L. and Singh, M. (2001). Recent developments in adjuvants for vaccines against infectious diseases. **Biomol. Engineering,** 18, 69-85.

Richter, L.R., Thanavala: Y., Arntzen, C.J. and Mason, H.S. (2000). Production of hepatitis B surface antigen in transgenic plants for oral immunization. **Nat. Biotechnol.** 18(11): 1167-1171.

Sandhu, J.S., Krasnyanski, S.F., Domier, L.L., Korban, S.S., Osadjan, M.D. and Duetow, D.E. (2000). Oral immunization of mice with transgenic tomato fruit expressing respiratory syncytial virus-F protein induces a systemic immune response. **Transgenic Res,** 9(2): 127-35.

Streatfield, S.J., Jilka, J.M., Hood, E.E., Turner, D.D., Bailey, M.R., Mayor, J.M., Woodard, S.L., Beifuss, K.K., Horn, M.E., Delaney, D.E., Tizard, I.R. and Howard, J.A. (2001). Plant-based vaccines: Unique advantages. **Vaccine** 19: 2742-2748.

Tacket, C.O. and Mason, H.S., (1999). A review of oral vaccination with transgenic vegetables. **Microbes Infect.** 1: 777-783.

Tacket, C.O., Mason, H.S., Losonsky, G., Estes, M.K., Levine, M.M., and Arntzen, C.J. (2000). Human immune responses to a novel Norwalk virus vaccine delivered in transgenic potatoes. **J. Infect. Dis.** 182, 302-305.

Tacket, C.O., Mason, H.S., Losonsky, G.,Clements, J.D., Levine, M.M., and Arntzen, C.J. (1998). Immunogenicity in humans of a recombinant bacterial antigen delivered in a transgenic potato. **Nat. Med.** 4(5): 607-609.

Thanavala, Y., Yang, Y.F., Lyons, P., Mason, H.S and Arntzen, C.J. (1995). Immunogenicity of transgenic plant derived hepatitis B surface antigen. **Proc. Natl. Acad. Sci.** USA. 92: 3358-3361.

Turpin, T.H., Reini, S.J., Charoenvit, Y., Hoffman, S.L., Fallarme, V. and Grill, L.K., (1995). Malarial epitopes express on the surface recombinant tobacco mosaic virus. **Biotechnology,** 13: 53-57.

Wigdorovitz, A.W., Carrillo, C., Dus Santos, M.J., Trono, K., Peralta, A., Gomez, M.C., Rios, R.D., Franzone, P.M., Sadir, A.M., Escribano J.M., Borca, M.V. (1999). Induction of protective antibody response to foot and mouth disease virus in mice following oral or parenteral immunization with alfalfa transgenic plants expressing the viral structural protein VPI. **Virology** 255: 347-353.

Yu, J. and Langridge, W.H.R (2001) A plant-based multicomponent vaccine protects mice from enteric diseases. **Nat. Biotechnol.,** 19(6):548-52.

13

Genetic Engineering as a Tool to Combat Biotic Stresses

Rakesh Tuli, P.K. Singh and Samir Sawant
National Botanical Research Institute, Rana Pratap Marg, Lucknow-226001

Precision breeding to improve field crops against biotic stresses needs greater att ention. A battery of potential pathogens: insects, fungi, bacteria, viruses, nematodes and weeds cause more than 30 percent annual loss to global crop productivity. Sophisticated mechanisms that perceive such threats and develop an adaptive response exist naturally in certain plant species and varieties. However, in several agronomically improved crop species, such sources of resistance are not available for exploitation through conventional breeding. A lot of good science for understanding plant-pathogen interaction and molecular basis of pathogenicity and resistance has been done through the last decade. Several novel approaches that interfere in the process of pathogenesis have been demonstrated. Impressive success has been achieved in commercial release of genetically engineered cotton, potato and corn cultivars for resistance to insects. Significant tolerance to a number of viruses has also been reported in commercially released transgenic cultivars of potato, papaya and squash. Weeds have been controlled in canola, corn, cotton, soybean and sugar beet by developing cultivars that selectively escape the treatment of a complimentary herbicide. However, results from molecular basis of pathogenesis and defence have given a variety of promising theoretical models.

The status of research at National Botanical Research Institute, Lucknow in the area of designing and deploying d-endotoxin genes for developing insect resistant cotton lines, designing artificial gene expression systems that may be strategically used against pathogens and progress with plant viruses is presented here. A general review of this fast moving field is provided to indicate several interesting directions of progress that may provide leads for strengthening defence mechanisms against biotic stresses in plants.

Protection Against Viruses

Since there are no effective chemical treatments against viral diseases, some of the earliest efforts were made in molecular genetics for plant defence against viral diseases. A variety of molecular approaches have been tried. Small size of viral genomes has facilitated the progress. Though molecular basis of the protection offered is not certain in several cases, coat protein mediated resistance, replicase and other viral genes, antisense RNA, movement protein genes, satellite RNA, post transcriptional silencing, use of antiviral proteins and antibodies have been reported to work with varialble levels of success.

Protection Against Fungal and Bacterial Pathogens

Most plants are successfully infected by only a small number of pathogens. Pathogen-specific mechanisms have evolved in several plant species, which have become non-hosts through millions of years of coexistence. They provide some of the natural models to understand the basis of plant defence. A powerful approach is the host resistance pathway, induced by virulence genes of the pathogen. Such interactions prevent pathogenesis in a variety of ways, including the production of phytoalexins toxic to the pathogen, PR proteins like enzymes that degrade fungal cell walls, deposition of barriers such as lignin etc. An avirulence gene may signal R gene mediated pathway, leading to hypersensitivity response, such as local necrosis. Ribosomal inhibitor proteins that inhibit protein synthesis in fungi by RNA N-glycosidase modification of 28S rRNA without inactivating 'self' ribosomes have been reported in some cases. A variety of peptides, called antimicrobial plant defensins have been reported from monocots, dicots, as also insects, mammals and other animals. Development of sustained resistance by deploying any of these strategies has not yet been reported.At least two pathways critical to mounting response against pathogens in plants have been identified. These are salicylic acid and ethylene / jasmonic acid induced pathways. Pathogen mediated activation of either of these induces resistance through a cascade of events which operate through a family of stress induced kinases. Reactive oxygen species is also suggested to initiate signal defence response. Overlapping steps among these pathways and even segments of cross talk in response to biotic and abiotic stresses have been reported using a number of mutants and gene expression profiles in *Arabidopsis*. A large number of pathogen-induced genes involved in metabolism, transport and transcription have been identified. Establishment of pathogenesis - specific essential pathways or proteins key to the infection process can provide vital clues for strategies for defence. For instance, the necrotrophic fungus *Sclerotinia sclerotiorumes* synthesises large amount of oxalic acid on infection. Mutants defective in oxalic acid biosynthesis are non-pathogenic. Oxalic acid enables the fungus to colonize the plant through its role in degradation of cell wall and interference in onset of plant defence. Transgenic plants over-expressing oxalic acid oxidase or oxalate decarboxylase confer resistance to tobacco and tomato. Therefore, improved understanding of the molecular basis of pathogenicity will be at the center stage of developing new targets and strategies to achieve defence response in crop plants.

Current Research at NBRI in Plant Defence Against Biotic Stresses

Protection against insects

A novel chimeric *CryIEC*-type d-endotoxin protein was designed at NBRI to cause complete mortality in a common lepidopteran pest, *Spodoptera* at all instar stages. This protein caused growth retardation or mortality of upto second instar larvae of *Helicoverpa* also. A gene coding for the novel d-endotoxin was designed for high level of expression in dicot plants and synthesised chemically. Two other synthetic genes, *CryIAc* and *CryIAa* that specifically target *Helicoverpa* were also synthesized. All three genes were designed and synthesized fully at NBRI. A patent on the first gene has been filed under PCT. The *CryIAc* is patented in other countries by Monsanto, St. Louism, USA. The other two genes designed and synthesized at NBRI have several modifications that may be functionally important. Transgenic cotton and tobacco lines expressing the novel d- endotoxin (*CryIE/C*-like) were developed. Coker cotton lines that express the d-endotoxin at 0.2 to 0.7 per cent of leaf protein show complete protection from *Spodoptera*. These are being back crossed with elite cotton cultivars in collaboration with Indian seed industry. Transgenic cotton with *CryIAc* is under development. The Spodoptera targeted gene is also being introduced into castor at Directorate of Oilseeds Research, Hyderabad under the APNL Biotechnology Programme. It is also being deployed for groundnut at the University of Hyderabad. Efforts to introduce *CryIAa, CryIF* and *CryIIA5* genes in other crops, including rice and pigeonpea are in progress under National Agricultural Technology Project.

Promoter Based Strategies for Plant Defence

A novel gene expression cassette, that gives a high level of constitutive expression of genes in dicot plants has been developed. Related patents have been filed in USA and Europe. Strategies that give expression at 1 to 4% of total protein in seed, root, stem and leaf were developed. Laboratory designed, synthetic promoters for unregulated and salicylic acid regulated expression in dicot plants were developed. These can be used for high level of expression of defence proteins in plants. Salicylic acid and wounding induced promoters can be deployed to express d-endotoxins. Tightly regulated pathogen specific promoters will be designed to express elicitor or signal proteins to initiate hypersensitivity or apoptosis type of local response in host plant, at the site of attack. Such promoters can provide powerful tools for developing pathogen specific defence strategies.

14

Potential Applications of Genetic Engineering in Improving Nutritional Quality of Oils in Crop Plants

Ram Rajasekharan
Department of Biochemistry, Indian Institute of Science, Bangalore - 560012.

Lipids are essential components of all living cells because they are obligate components of biological membranes and energy reserves. Oils and fats are glycerol esters of fatty acids (triacylglycerols) and are mainly derived from plant and animal sources. Vegetable oils are the major source of edible lipids, accounting for more than 75% of the total fats consumed across the world. About 75% of the oils extracted, are from the endosperrm of oilseeds, like soybean, peanuts or oilseed rape, whereas the remaining 25% are produced from the pericarp of oil fruits, mainly oil palm and olive. The oil content of seeds of different plant species varies from 4% to over 60% of dry weight. Over 95% of the storage lipids in the seeds are present as triacylglycerol. The global demand for plant oils has intensified our efforts to genetically modify oilseeds mainly to enhance oil yield, nutritional quality and to generate novel oils with desired fatty acid composition. Rational approaches with an emphasis on the above, demand a detailed knowledge of the biochemistry and molecular biology of seed oil formation.

Biosynthesis of Triacylglycerol

Triacylglycerol (TAG) biosynthesis is shown to involve four reactions: acylation of glycerol-3-phosphate (G3P) to Iysophosphatidic acid (LPA); acylation of LPA to phosphatidic acid (PA); hydrolysis of the phosphate ester bond of PA, yielding diacylglycerol (DAG); finally, acylation of DAG to TAG. Endoplasmic reticulum has been considered to be the only site for TAG biosynthesis.

In the oleaginous yeast (*Rhodotorula glutinis*), TAG biosynthetic machinery is mainly localized in the cytosol. The purified 10 S multienzyme triacylglycerol biosynthetic complex (TBC) consisted of LPA acyltransferase, PA phosphatase, DAG acyltransferase, acyl-acyl carrier protein synthetase, acyl carrier protein (ACP) and superoxide dismutase. (Gangar *et al.*, (2001) *J. Bioi. Chern.* **276**, 10290-10298; Gangar et al., (2001) *Biochem. J.* **260**, 472-479). The current opinion in the literature is that, TAG biosynthesis occurs in the microsomal membranes and fatty acyl-CoA esters are the substrates for the acyltransferases. Our results provide the first direct evidence for the existence of a soluble triacylglycerol biosynthetic complex in oleaginous yeast that contains acyl-ACP synthetase and ACP for the activation of fatty acids. A function based interactive approach was employed for identifying the above components of the TAG biosynthesis in *R. glutinis*. The newly identified pathway is responsible for TAG accumulation in the oleaginous yeast (Gangar *et al.*, (2002) *Biochem. J.* **365**, 577-589). To expand and to complement the above finding, we used oilseeds as another model system to study TAG biosynthesis.

Monoacylglycerol pathway in oilseeds

In plants, *de novo* biosynthesis of triacylglycerol has been shown to occur in microsomal membranes by the sequential acylation as described earlier. The soluble fraction of immature peanut (*Arachis hypogaea*) was capable of dephosphorylating [3H] LPA generate monoacylglycerol (MAG). The enzyme, LPA phosphatase, was purified to apparent homogeneity from developing peanuts (Shekar *et al.*, (2002) *Plant Physiol.* **128**, 988-996). DAG formation was catalyzed by MAG acyltransferase (EC 2.3.1.22) that transferred acyl moiety from acyl-CoA to MAG. This enzyme was purified to 6,608-fold with the final specific activity of 15.9 nmol min^{-1} mg^{-1} (Tumaney *et al.*, (2001) *J. Bioi. Chem*, **276**, 10847 - 10852). The final acylation step in TAG formation is shown to be catalyzed by soluble DAG acyltransferase and this enzyme has been overexpressed in *E. coli*. The transformant is capable of producing triacylglycerol. Our results provide the first evidence for the presence of an alternative MAG pathway for triacylglycerol synthesis in developing oilseeds.

The isolated new genes will be used to alter the nutritional quality of the needs.

Triacylglycerol Hydrolysis

*A New Thermally Stable Alkaline Lipase from Rice Bran**

Rice (*Oryza sativa*) bran oil is typically an oleic-linoleic type fatty acid, and its physical -chemical properties qualify it for good quality edible oil. However, complete utilization of bran oil suffers from the fact that there is large accumulation of free fatty acids (FFA), which has been attributed to the presence of lipase activity. Identifying and characterizing the lipases from bran is essential to device efficient methods to overcome the problem of rice bran oil instability.

A thermally stable lipase was first identified in rice (*Oryza sativa*) bran, and the enzyme was purified to homogeneity using Octyl-Sepharose chromatography. The enzyme was purified to 7.6-fold with the final specific activity of 0.38 μmol min^{-1} mg^{-1} at 80 °C using (9, 10 - 3H) triolein as substrate. The purified enzyme

124

was found to be a glycoprotein of 9.4 kDa. Enzyme showed a maximum activity at 80 °C and at pH 11.0. The protein was biologically active and retained most of its secondary structure even at 90 °C as judged by the enzymatic assays and far-ultraviolet circular dichroism spectroscopy, respectively. Differential scanning calorimetric studies indicated that the transition temperature was 76 °C and enthalpy 1.3×10^5 Calorie mol^{-1} at this temperature. The purified lipase also exhibited phospholipase A$_2$ activity. Colocalization of both the hydrolytic activities in reverse-phase high-performance liquid chromatography and isoelectric focusing showed that the dual activity was associated with a single protein. Apparent K_m for triolein (6.71 mM) was higher than that for PC (1.02 mM). The enzyme preferentially hydrolyzed the *sn-2* position of PC, whereas it apparently exhibited no positional specificity toward triacylglycerol. This enzyme is a new member from plants in the family of lipases capable of hydrolyzing phospholipids.

The identified lipase is under commercial production to use in various industrial applications.

15

Risk Assessment of Genetically Modified Foods in the Indian Context

S. Vasanthi and Ramesh V Bhat
Food and Drug Toxicology Research Centre
National Institute of Nutrition, Hyderabad - 500 007

In recent years due to globalization of trade, there is an increased movement of food commodities across transnational borders, which is giving rise to newer food safety concerns. The major aspect of this concern which is emerging as the most controversial is the use of modern biotechnology for the genetic modification of plants, microorganisms and animals for the production and processing of food.

GM foods are made through recombinant DNA technology where genes coding for specific properties and characteristics are made to become an integral part of the transgenic plant genome. The cloned gene can originate from different sources like plant, microbes, animal or even synthetic DNA sequences designed to encode a specific function. The gene is transferred in the form of a construct, which contains various functional elements like promoters and terminators that are required for the gene to express/function (*Robinson, Scott and Gackle 2000*).

Genetically engineered foods originate from variety of plants, animals and microorganisms as a result of gene transfer technology. These include major crops such as rice, wheat, maize, cotton, soybean, sunflower, barley and mustard, microorganisms like lactic acid bacteria and animals like sheep and fish. In India, research on transgenic plants/crops namely cotton, rice, mustard, potato, vegetables are in progress at several organizations comprising government, universities and private laboratories. The GM crops to be considered in the Indian context include Bt (*Bacillus thuringiensis*) cotton, the commercial cultivation of which has already been approved, and crops awaiting trials for commercialization like hybrid mustard and those at the research stage in the laboratory and field like disease and insect resistance vegetables such as brinjal, tomatoes, cauliflower and sugarcane and crops conferring nutritional benefits like rice and potatoes.

Through such technology various genes have been transferred to plant to confer properties like improved product quality (delayed ripening of fruits, nutritive and processing value), pest resistance (insects, nematodes and viruses) and agronomic benefits (herbicide resistance, environmental stress) (*Mazur, Krebbers and Tingey 1999; Ye, et al, 2000; Kleter et al, 2001*).

The introduction of genetically modified foods (GM) has generated much debate worldwide in view of the uncertainties regarding safety of these foods. The potential for the recombinant gene inserted as well as its products to cause health and environmental risks is as yet unclear, as, (i) scientific evidence regarding their toxicity or health risks is limited/inadequate, (ii) the methodology that is currently used for assessing such risks is not robust or sensitive enough and (iii) the molecular/ genetic effects of the process of genetic engineering are unpredictable in nature (*Pusztai 2001; Kuiper, Noteburn and Peijenburg 1999*). The National Academy of Sciences of various countries recommended that potential adverse effects and long-term health effects of GM foods be identified and monitored (*Anon 2001*). They also stressed the importance of risk assessment of GM foods to ensure the safety of these foods.

Risk assessment of GM Foods

Risk assessment includes a safety assessment, which is designed to identify whether a hazard, nutritional or any other safety concern is present, and, if present, to gather information on its nature and severity. The safety assessment is based on the determination of similarities and differences between the GM food and its traditional counterpart. Factors taken into account include both intended and unintended effects of GM, identification of new or altered hazards and identification of changes relevant to human health in key nutrients. The risks and concerns expressed on the safety of GM foods led many countries and international organizations to establish risk assessment procedures and labeling requirements for GM foods.

International fora addressing the risk assessment of GM foods

Risk assessment of GM foods has been addressed by several international organizations like Food and Agriculture Organization (FAO), World Health Organization (WHO), Organization for Economic Cooperation and Development (OECD) and the Codex Alimentarius Commission (CAC) and also in multilateral agreements like the World Trade Organization (WTO) and the Cartegena Protocol on Biosafety which are involved in issues relating to trade in GMOs.

The Joint FAO/WHO consultations on foods derived from biotechnology

Various consultations have been convened by the FAO and WHO and the OECD identified the concept of substantial equivalence as a tool for assessing the safety of GM foods (OECD 1993; WHO 1995). It involves a comparative approach where the relative safety of GM food or food component to an existing food or food component is established. Safety is established by the demonstration that there is no significant difference in a range of phenotypic and compositional characteristics like nutrients, toxicants, anti-nutrients, agronomic traits, etc. between the GM and non-GM food. The outcome of the comparison may be that the GM food is substantially equivalent, except for the inserted trait or not equivalent at all. Depending on the outcome the foods are subjected to further toxicological studies.

The Codex Alimentarius Commission of the FAO/WHO set up an *Ad hoc* intergovernmental task force on foods derived from biotechnology to develop standards, guidelines, recommendations for foods derived from biotechnology. The task force identified that the risk assessment of GM foods requires scientific data which addresses the current safety concerns of GM foods like the effect of the genetic modification process including the function and properties of newly inserted genes, the safety and nutritional properties of newly expressed substances in the food and their impact on diet, potential for allergenicity, potential for gene transfer to human and animal cells, and unexpected changes in the composition of the modified product due to insertion of novel genes or suppression of constituent genes. A draft guideline for the conduct of safety assessment of foods derived from genetically modified plants has been brought out by the task force (Codex Alimentarius Commission 2002).

Cartegena Protocol on biosafety

The Cartagena Protocol was negotiated under the auspices of the Convention on Biological Diversity (CBD) in 1992 (Anon 2000). The Protocol provides rules for safe transfer, handling and disposal of Living Modified Organisms (LMOs) or Genetically Modified Organisms (GMOs). Its aim is to address the threats posed by living modified organisms (LMOs) to biological diversity, also taking into account the risks to human health. The protocol takes into account the general principles of risk assessment developed by international bodies.

Two features of the protocol, the Advanced Informed Agreement (AIA) and the Precautionary Approach are being incorporated in risk/safety assessment procedures in many countries particularly in the context of trade in GMOs. The AIA provides for a prior assessment by importing country of GMOs intentionally introduced into the environment like seeds for plantation, live fish for release etc. This agreement calls for documentation and identifaction of LMOs which include the relevant trait, information handling, storage, transport and use along with a full report or risk assessment. In making the decision to import, the Protocol allows a precautionary approach to be used to restrict or ban the GMO if there is a lack of scientific certainty due to insufficient information on the potential risks that LMOs can have on biodiversity and human health.

WTO Agreements

The WTO is mainly involved in establishing rules for international trade in GM foods (*Zarrilli 2000*). Two agreements in the WTO apply to risk assessment and labeling of GM foods. These are the Agreements on Sanitary and Phytosanitary Measures (SPS) and Technical Barriers to Trade (TBT). The risk assessment of GM foods for trade requirements is addressed under the agreement on Sanitary and Phytosanitary Measures (SPS). This agreement deals with application of food safety and animal and plant health regulations. By imposing science-based disciplines and requiring risk assessment based on science and applied only to the extent necessary to protect human, animal or plant life or health, it aims to prevent governments from using health and safety laws to limit international trade. The TBT agreement assists to ensure that a WTO member does not use domestic regulations, standards, testing and certification procedures to create unnecessary obstacles to trade. It encourages countries to use international standards where appropriate.

Strategies for risk assessment of GM foods in developed countries

The risks and uncertainties surrounding the process of genetic engineering and resulting GM products have prompted many countries to regulate the development and use of GM foods.

In most developed countries like USA and the EU risk assessment of GM foods is based on the determination of substantial equivalence. However the use of this approach for regulating GM foods differed considerably between the two countries.

In the USA the Food and Drug Administration (FDA) is the main agency for GM food safety considerations (*Neumann 1999*). According to FDA the GM foods are not substantially different from their traditional counterparts and these foods are safe for consumption. The US is also opting for narrowing the scope of the Cartagena Protocol so as to include the AIA only for GMOs intended for direct release into environment and limit the use of precautionary principle in decision making. The US argues that decision on biosafety must be based on scientific evidence and the precautionary principle is inconsistent with other multilateral agreements on GMOs particularly the SPS Agreement of the WTO. In the EU however, GM foods are considered to be fundamentally different from conventional foods owing to the presence of the inserted transgene and its products. Thus GM foods undergo a rigorous risk assessment protocols before being approved for marketing and consumption (Official Journal of the European Communities 1997). The risk assessment takes into account the identification of any characteristic which may have potential adverse effects either direct or indirect, immediate or delayed. Special attention is focused on the method of development of GMO and also examine the associated risks of the gene products and possibility of gene transfer. Recently the Council has proposed mandatory monitoring requirements of long term effects associated with interactions with other GMOs and the environment. The EU is also supportive of a strong CBD protocol that takes into consideration risks to human health of all the GM commodities besides those meant for release into environment, and the precautionary principle.

Recently the EU also introduced mandatory monitoring requirements of long term effects associated with interaction of GM foods with other genetically modified organisms and the environment (Official Journal of the European Communities 2001).

Risk assessment of GM foods in the Indian context

In India the existing framework for risk assessment of GM foods is primarily focused at the environment level and R&D stage. The Ministry of Environment and Forests of the Government of India through its Gazette Notification of the Environment Protection Act 1986, set up the rules and procedures for handling GMOs and hazardous organisms mainly from the environmental angle (Ministry of Environment and Forests 1989). Apart from this Act, various Committees have been established to deal mainly with recombinant research, use and application of GM foods (DBT 1998). These are the Genetic Engineering Approval Committee (GEAC), the Review Committee for Genetic Modifications (RCGM), the Recombinant DNA Advisory Committee (RDAC), and the Institutional Biosafety Committee (ISBC) (DBT 1998). However, risk assessment framework for GM foods as part of food safety regulations is yet to be established in India. In this context, the Ministry of Health, Government of India has set up an expert committee to work out the risk assessment protocols to form as a basis for regulation of GM food safety.

The risk assessment of GM foods from the point of view of food safety in the Indian context need to address the following issues: (i) effect of the GM process, (ii) the safety and toxicity of the gene products, (iii) potential for allergenicity, (iv) nutritional implications and (v) occurrence and implications of unintended effects.

The genetic modification process

The use of recombinant DNA (rDNA) technology in the production of GM foods involves transfer of genes from different species into the food producing organism. Such a transfer is facilitated along with various regulatory elements that are required for the gene of interest to express in the host organism. Very often these components like promoters, terminators and enhancers are obtained from bacterial or viral sources. The function and properties of such components to have any impact on the safety is not yet clearly known as they have (i) the potential to induce toxicity, (ii) transfer to gut flora or (iii) produce unintended effects by changing the level of transgene expression, modification or disruption of functional genes of host plant at the site of insertion or give rise to pleiotropic effects (*Kuiper et al, 2001; Pusztai 2001; Butler and Reichardt 1999*). In order to assess risk it is important to assess and characterize the genetic modification process including the donor organism, vector used and its characteristics, the genetic construct and its components, as well as the type and nature of the gene products.

Safety of the new gene products

Recombinant DNA techniques enable the introduction of DNA that can result in the synthesis of new substances in the plants. The gene product may be the final active product or act as an enzyme or hormone that mediate the production of other compounds. The new components may be conventional components of plant foods like proteins, fats, carbohydrates, vitamins which are novel in the context of the recombinant DNA plant. New substances may also include new metabolites resulting from the activity of enzymes generated by the expression of the introduced DNA. Depending upon the modified component, the GM food may contain or consists of GMO or produced from GMO but not contain the GMO.

The potential toxicological and other compositional changes arising from the gene products that could be harmful to human health are considered important for risk assessment of GM foods. Various toxicants are known to be inherently present in different plants. Genetic modification has the potential to alter such constituents or produce newer toxicants.

The chemical nature and function of the newly expressed substances, the level and site of expression of the transgene in different parts of the plant particularly in the edible part of the recombinant DNA plant and current dietary exposure and possible effects of such modifications on population sub-groups are important considerations for assessing the safety of the gene products. GM crops that have been developed for insect resistance like cotton, potato and maize contain pesticidal proteins. Such foods need to be subjected to various toxicological studies in experimental animals which include short term oral toxicity and sub-chronic toxicity studies besides assessing the nature and function of the protein (EPA 2000).

Nutritional considerations

The information of rDNA techniques provided a potential to produce food with improved nutritional value. These changes are brought about by altered nutrient composition and levels or change in the functionality of the product. Currently transgenic plants with improved nutritive value include GM rice with enriched vitamin A and GM soybean and rapeseed with modified fatty acid. The impact of such intended modification in nutrient levels in crop plants can affect nutritional status of the individual. On the other hand, there exist a potential for unexpected alteration in nutrient, which can affect nutrient profiles of the product as well as nutritional status of people. Examples of such unintended alteration include increase in prolamines during genetic modification in rice resulted in a decrease in glutelin levels, introduction of β-carotene in rice led to accumulation of xanthophylls (WHO 2001 a). Such changes assume relevance particularly in India when staple foods like rice and maize which are consumed in large quantities are genetically modified. In addition to these foods the impact of processed foods containing GM ingredients on human health also needs to be considered.

Potential allergenicity of new proteins

Most traits introduced into the crop result from the expression of one or more proteins. Currently the allergenicity potential of GM foods is assuming much significance in the risk assessment of GM foods (WHO 2001 b). GM foods of particular concern are those that have been modified for insect resistance. Some of the insecticidal proteins have been shown to possess allergenic properties. An example is the unexpected occurrence of StarLink maize that is genetically modified to produce the *Cry9C* protein for insecticidal properties in the food supply (*Anon 2000 c*). This protein has been shown to possess allergenic properties and was approved only for animal feed purposes (*Fox 2001*).

Assessment of allergenicity focuses on the source of the gene which can be allergenic one or of unknown allergenicity. Information on the sequence of homology of the newly introduced protein to known allergens, *in vitro* and *in vivo* immunologic assays and assessment of key physico-chemical properties of the newly introduced protein like stability to proteolytic and acidic conditions of the digestive tract need to be taken into account.

When the gene is from a crop of known allergenicity it is easy to establish whether the GM food is allergenic using *in vitro* tests such as RAST or immunoblotting. This test has been used to assess the allergenicity of GM soy expressing brazil nut 2S storage protein that is shown to have allergenic properties. However, it is often difficult to establish allergic potential of the GM food which has a (i) transgene transferred from a source that is not eaten before; (ii) with unknown allergenicity; (iii) when expression of a minor allergen is increased due to genetic modification (*Lehrer, Horner and Reese 1996; Kuiper et ai, 2001*). Although comparison with amino acid sequence homology of known allergens is used to assess unknown allergenicity, these are indirect methods which cannot establish with certainty the allergenic potential of the GM crop (*Pusztai 2001*).

Additional tests are currently being proposed to assess allergic potential of GM food which take into account the level and site of expression and function of the novel protein (WHO 2001 b). This factor becomes important when such proteins, are expressed in edible portion that is consumed.

Potential for gene transfer to human and animal cells

Concern has been expressed on the possibility of transfer of GM DNA in plant to microbe and mammalian cells. Safety issues have been focussed on marker genes particularly the herbicide resistant gene and antibiotic resistance genes (WHO 1993). Currently, the use of alternative transformation system that do not use antibiotic selection markers is being considered. The specific issue related to the use of antibiotic marker genes expressed in the plant is the potential to adversely effect the therapeutic efficacy of orally administered antibiotics. Factors that should be considered in the assessment of potential impact of such antibiotic efficacy include, clinical use of the antibiotic, the potential for compromising the therapeutic efficacy of the orally administered antibiotic and the safety of the gene product.

Occurrence and implications of unintended effects

Occurrence of unintended effects of genetic modification may be in the form of acquisition of new traits or loss of existing traits. These effects may arise from, (i) the random insertion of DNA sequences into the plant genome resulting in disruption or silencing of existing genes; (ii) formation of new or changed patterns of metabolites expressing enzymes which may be expressed at high levels and give rise to secondary biochemical effects (*Codex Alimentarius Commission 2002; Firn and Jones 1999*).

Many unintended effects are largely predictable on the basis of knowledge of the place of transgenic insertion, the function of the inserted trait or its involvement in metabolic pathways. Other effects are unpredictable due to limited knowledge of gene regulation and gene-gene interactions or pleiotropic effects.

Molecular, biological and biochemical techniques can analyse potential changes due to unintended effects of genetic modification. The assessment of unintended effects should take into account, (i) the agronomic/phenotypic characteristics of the plant; (ii) molecular characterization including stability of introduced DNA; (iii) chemical analysis of key nutrients, anti-nutrients, toxicants, vitamins, minerals, and other compounds typical of the plant; (iv) alteration of the metabolites; (v) any effects due to food processing (*Kuiper et al 2002*). Profiling techniques like proteomics, DNA microarray and metabolomics are being developed which allow identification of potential changes in the modified organism at the DNA level, during gene expression and protein translation and also in various metabolic pathways (*Kuiper, Kok and Noteborn 2000*).

Conclusions

A wide variety of GM crops such as saline, drought, herbicide tolerant, insect, bacteria, viral or fungal resistant and those conferring nutritional benefits are being sought to be introduced in India. In view of this there is an urgent need to establish scientifically based guidelines and protocols for the risk assessment of GM foods in India. Such guidelines and their implementation are also essential in view of the possibility of several unauthorized varieties entering the food chain.

References

Anon (2000) ConAgra foods recalls certain meal, flour, grits and cereal products. **http://www.safetvalerts.com**

Anon (2000) Cartagena Protocol on Biosafety to the Convention on Biological Diversity. Secretariat of the Convention on Biological Diversity, Montreal.

Anon. (2000) Scientists back GM for Third World. **Nature** : 406:115.

Butler D, Reichardt, J. (1999) Long-term effects of GM crops serves up food for thought. **Nature** 398 : 651-656.

Codex Alimentarius Commission (2002) Draft report of the third session of the Codex Ad Hoc Intergovemmental Task Force on Foods derived from Biotechnology. CAC Alinorm 03/34. FAO, Rome.

DBT (1998) Revised guidelines for research in transgenic plants and guidelines for toxicity and allergenicity evaluation of transgenic seeds, plants and plant parts. Department of Biotechnology, Ministry of Science and Technology Government of India.

EPA (2000) *Bt* Plant - pesticides Biopesticides Registration Action Document. Preliminary risks and benefits Sections. *Bacillus thuringiensis* Plant pesticides. US Environment Protection Agency. Office of the Pesticide Programmes. Biopesticides and Pollution Prevention Division. Environmental Protection Agency, USA.

Fim RD and Jones CG (1999) Secondary metabolism and risks of GMOs **Nature** 400:13-14.

Fox JL (2001) EPA reevaluates StarLink license. **Nature Biotech** 19:11

Kleter GA et al, (2001) Regulation and exploitation of genetically modified crops. **Nature Biotech** 19: 1105-1110.

Kuiper HA, et al (2001). Assessment of the food safety issues related to genetically modified foods. **The Plant J.** 27 : 503-528.

Kuiper HA, et al (2002) Safety aspects of novel foods. **Food Res Int** 35 : 267-271.

Kuiper HA, Kok E, Noteborn JPM (2000). Profiling techniques to identify differences between foods derived from biotechnology and their counterparts. Joint FAO/WHO Expert Consultation on Foods Derived from Biotechnology.

Kuiper HA, Noteborn and Peijnenburg ACM (1999). Adequacy of methods for testing the safety of genetically modified foods. **Lancet** 354 .1315-1316.

Lehrer SB, Homer WE, Reese G (1996) Why are some proteins allergenic ? Implications for biotechnology. **Crit. Rev. Food Sci. Nutr.** 36: 553-564.

Mazur B, Krebbers E, Tingey S (1999) Gene discovery and product development for grain quality traits. **Science** 285 : 372-375.

Ministry of Environment and Forests (1989) Rules for the manufacture, use, import, export and storage of hazardous microorganisms/genetically engineered organisms or cells. Issued by the Union Ministry of Environment and Forests, Government of India vide Notification No. G.S.R. 1037(E) dated 5th December 1989. MOEF, New Delhi, India.

Neumann DA (1999) Safety assessment and regulation of genetically modified foods in North America. International Conference on Biotechnology for Sustained Productivity in Agriculture. ILSI (India) New Delhi 110016 pp IV-A-2-IV-A-2-6

OECD (1993) Safety evaluation of foods derived by modern biotechnology: Concepts and Principles. Organization for Economic Cooperation and Development, Paris, France.

Official Journal of the European Communities L106/1 2001. Directive 2001/I8/EC of the European Parliament and of the Council on the deliberate release into the environment of GMOs and repaling Council Directive 90/220/EC.

Official Journal of the European Communities, (1997), L 253 Vol. 40.

Pusztai, A (2001) Genetically modified foods : Are they a risk to human/animal health? http://www.actionbioscience.org/biotech/pusztai.html

Robinson SR., Scott N, Gackle A (2000) Gene technology and future trends. **Asia Pacific J. Clin. Nutr.** 9: S113-S118.

WHO (1993) Health aspects of marker genes in genetically modified plants. Report of a WHO workshop. WHO-Food Safety Unit, Geneva.

WHO (1995) Application of the principles of the substantial equivalence to the safety evaluation of foods or food components from plants derived by modern biotechnology. Report of a WHO workshop. WHO-Food safety Unit, Geneva.

WHO (2001a) Safety assessment of foods derived from genetically modified microorganisms. Report of the Joint FAO/WHO Expert Consultation on foods derived from biotechnology. World Health Organization. Geneva.

WHO (200lb) Assessment of possible allergenicity (proteins). Joint FAO/WHO Consultation on foods derived from biotechnology, Rome.

Ye X, et al, (2000) Engineering the provitamin A (β-Carotene) biosynthetic pathway into (carotenoid-free) rice endosperm. **Science** 287: 303-305.

Zarrilli, S., International trade in genetically modified organisms and multilateral negotiations. A new dilemma for developing Countries. United Nations Conference on Trade and Development (UNCTAD), 2000. **http://www.unctad.org:/en/docs/poditetncddl.en.pdf**

16

Risk Assessment Studies on Genetically Modified Crops

Arvind Kapur
Nunhems Seeds Pvt. Ltd.
Dhumaspur Road, Badshahpur, Gurgaon - 122 001 (India)

The sequencing of human genome, rice genome and many other organisms has opened up many vistas in our understanding of intricacies of living organisms on the earth. It is now possible to transfer genes not only within the related genera and species but also among different kingdoms. The major issues related to these technologies are their safety relating to environment and food. The major debate on the use of GM technology in agriculture, food and the environment, is lack of knowledge on possible impacts and their remedial measures. Research results can resolve uncertainties and provide a sound basis for risk assessment and science based regulation through pre-normative research and lead to the establishment of best practice in a constantly evolving way.

Biosafety research over the past quarter century has played a key role in the development and diffusion of modern biotechnology products and applications and health care, agro food and environment. The benefits of the more precise methods are becoming clear but as always with innovations, the precautionary approach demands that uncertainties and conjectural risks may be addressed by corresponding research. The results of research and growing practical experience feeding into regulatory and risk management policies have enabled these to be regularly adapted to facilitate safe innovation and contributing to the excellent safety records. This will install public confidence in the technology and its products.

Historically ignorance has been a major driver for apparently irrational and backward looking behaviour. For the simple reason that most humans feel uncomfortable when confronted with things and issues they do not understand. It is so much simpler to condemn something than to attempt to understand it. People do applaud the change but it carries within it the problems we face in deciding how to communicate the risk / benefit analysis of the use of new technologies.

Risk Identification

It is important first to identify the potential risk related to modern technologies and then research can be directed to diminish or eliminate its impact which is a key part of risk management. The example is virus resistance. In this case a viral coat protein gene is expressed in transgenic plant which could have an impact on the epidemiology of viral diseases via changes in the interactions with virus vectors such as insects, nematodes or fungi. The researchers showed that it is possible to modify the transgene to eliminate the interaction with the vector while maintaining its ability to confer virus resistance. This is how the approach should eliminate the source of potential risk. Another similar problem solving mode is to develop a system for the selection of transformed cells that would not be based on antibiotic resistance. Although no risks have been scientifically associated with the antibiotic resistance genes commonly used to generate transgenic plants, the new system would avoid this highly volatile problem of development of antibiotic resistance in humanbeings and could thus contribute to improving public acceptance of transgenics.

The new traits which are encoded by single gene whose products have little or no effect on the plant biochemistry is easy to handle. As more complex traits that profoundly modify plant biochemistry are introduced into crops such as ones that modify the characteristics of important plant products (oil, carbohydrate, etc.) or those that confer resistance to abiotic stresses (salinity, cold, drought, etc.) much more complex ecological questions will be raised which will of course require further and more sophisticated biosafety evaluation.

Risk Assessment

Gene transfer between microorganisms and plants

The release of GMO into the environment has led to increased interest in possible interactions that may occur between resident organisms. The trans-kingdom gene transfer from bacteria to eukaryotes has been demonstrated in the laboratory. In the absence of documented proof of genetic transfer from plants to bacteria research experiments have to be designed to assess possible gene transfer between plants and microorganisms. Research experiments using transgenic plants containing gentamycine resistance gene and luciferase gene conferring bioluninescence to plants and bacteria have been designed. The soil bacterium, *Agrobacterium tumefaciens* was used as a biological tool to detect plant / bacteria DNA transfer. No evidence for gene transfer from plants to bacterium in a plant tumour was obtained.

Horizontal gene transfer between organisms

Horizontal gene transfer is the movement of genetic information between species. There is only circumstantial evidence available for horizontal transfer between different eukaryotic organisms.

Tackling food safety concerns of GMO's

The major focus of GM Foods has been their safety as regards food use and the environment. Consequently, analysis of new evaluation techniques and protocols is important to establish safety of transgenic foods. Every country has food laws where food for human and animal consumption should have a safety level as recommended by the law. Many traditional foods have been introduced in different countries from other countries without the testing which is now applied strictly to GM foods. Some of these issues concern traceability of genetically modified material, potential for transfer of the introduced gene to other species, safety of the introduced gene products including allergenicity and the question of substantial equivalence. Another issue related to this is to examine whether the DNA can be taken up and incorporated into the genome of intestinal microorganisms. If this is demonstrated, the implication of this DNA transfer will need to be assessed in terms of potential impact on intestinal flora and host interactions.

New methods for the safety testing of transgenic food

The European Union has developed a methodology called Safotest which has been recommended to ensure safety of GM foods. Based on OECD different animal studies have been designed to ensure perfect assessment for the safety of food. The objectives of these studies are as follows:

- The nutritional and toxicological consequences of inserted gene.
- The potential of pleiotropic effects in the host organism due to insertion.
- The allergenicity of expressed proteins and novel food stuffs.
- The potential of gene transfer to human and animal gut flora.

The principle of substantial equivalence has been established which include a hierarchy of comparisons including the choice of comparators and appropriate statistical analysis.

Consumer attitude and decision making with regard to GMO

Public understanding and opinion are extremely important factors in the process of the integration of genetically modified crops. Many surveys showed that to a large degree, perception of benefits was determined by risk associated with using GM in food. There is a relationship between prior attitude and choice behaviour, regardless of the kind of information or information source.

Case Study on GM Mustard *(Brassica juncea)*

In 1994 Proagro Seed Co. Ltd. (PSCL) has established a joint venture company with Plant Genetic Systems (PGS), Belgium to develop transgenic technology in Mustard *(B. juncea)* and in vegetables. In Indian mustard the yields of released varieties was constant for the last 10 years at around 1.5 ton / ha. The hybridization system (CMS) (GMS) were not stabilized though a lot of research was conducted in the ICAR system and in universities. Based on that experience a transgenic technology of Plant Genetic Systems called Seedlink™ which was already tested in Canada in *B. napus* was adopted. The PGS has launched commercial hybrids of *B. napus* in Canada successfully and enhanced the yield upto 30%. Based on the experience, experiments were started for converting *B. juncea* lines with the help of the transgenes Barnase, Barstar & Bar.

Expression of the genes

Barnase : expression in the flower only during a specific stage of floral development (expresses in the transient tapetum tissue only).

Barstar : expression in the flower only during a specific stage of floral development (expresses in the transient taptum tissue only)

Bar : expression limited to green tissue (leaf, stem) only. No (or negligible) presence in root and seed.

After completing the greenhouse trials and stabilizing the hybridization system in mustard, limited field study started for environmental and food safety as per the DBT guidelines. Under the environmental safety studies pollen flow studies for 3 years out crossing to related species which exists in the Indian agricultural system were completed. The invasiveness/ competitiveness/ weediness studies, effects on soil microflora to demonstrate the impact of transgene have also been completed.

Environmental Safety

- Pollen Flow
- Outcrossing to related species
- Invasiveness / Competitiveness / Weediness
- Soil microflora

Under the agronomic performance the company has completed the initial hybrid trials and advance hybrid trials from 1997 to 2000. In 1999-2000 the company has also completed the multilocational trials. All these results indicate the positive outcome and the transgenic was compared with the non-transgenics in all these studies and no significant deviations were reported from these studies.

Agronomic Performance

- Yield evaluation
- Stability of yield

Barnase - Barstar technology

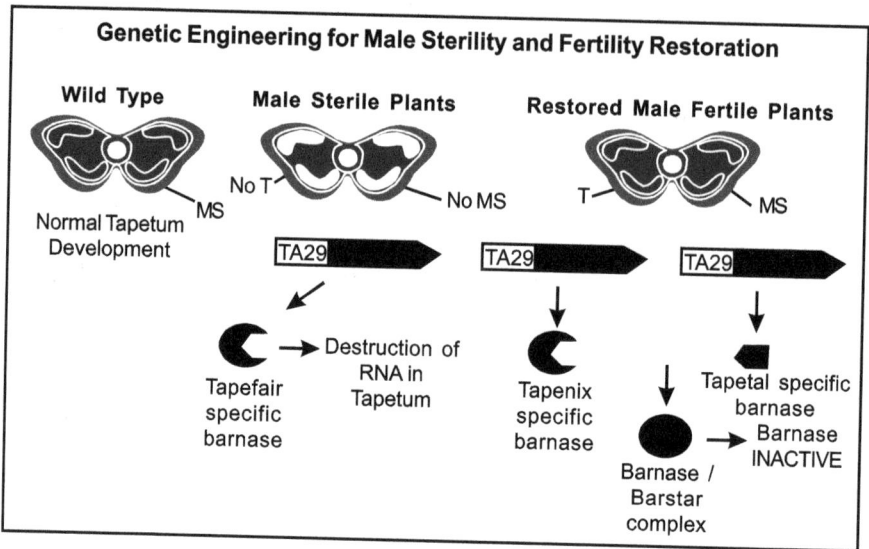

Genetic Engineering for Male Sterility and Fertility Restoration

Under the food and feed safety studies the company has completed compositional analysis (substantial equivalence), allergenicity and toxicological studies.

Food and Feed Safety (Substantial equivalence)

- Compositional (nutritional) analyses
- Allergenicity
- Toxicological studies

All these studies established the substantial equivalence with the non-transgenic counterparts and the studies reported no allergenicity and toxicity. The company also has completed the molecular characterization of the product and whole package was submitted to RCGM.

The study of Indian mustard and the development of hybrids with Seedlink™ system has proved that this technology is very stable and can give the hybrids which yield higher than the present varieties.

17

Building Capabilities in Biotechnology in India
A Forecasting Exercise for 2010

S. Visalakshi
*National Institute of Science Technology in Development Studies,
Dr, K.S. Krishnan Marg, New Delhi - 110012, India.
email : visha_3@yahoo.com*

Introduction

By now it is understood beyond doubt that commercialisation and industrial competitiveness in Biotechnology (BT) depends upon adequately trained scientific and technical personnel. The degree of sophistication of R & D personnel with respect to the state of the art in biotechnology will be a major factor determining the success of the companies attempting to commercialise biotechnology. High technology firms of the USA have consistently ranked 'quality of education' and 'availability of skilled work force' as among the most crucial elements for success in BT commercialisation [1, 2]. As per Gelhart and Cass [5] the accessibility to educational institutions determines even the location of the BT firms.

It is also, by now, clear that biotechnology is not a single discipline but an inter-disciplinary activity applying principles of several disciplines of science and engineering for production of goods and services [3]. Hence the type of manpower needed for achieving successful commercialisation is not dependent on skills in biotechnology as such - but abilities, knowledge and skills in a variety of disciplines contributing to it such as - microbiology, molecular biology, immunology; genetics, cell biology, tissue culture, biochemical and chemical engineering, computer modeling, etc. Any limitation in certain skills has led to weaknesses in various countries at the level of R & D leading to varied competitiveness. If one goes through the available literature on the status of human resource involved in BT in different countries [4], it clearly draws the picture of strengths and weaknesses of that country in its ability to produce BT - based products and its competitiveness.

Understanding of the need and importance of manpower at various levels for successful commercialisation of BT makes it necessary to study the patterns of deployment of manpower in the present situation, and assess the needs for the future and plan for development of required numbers of personnel with requisite specified capabilities. This would greatly avoid delays and failures of BT commercial ventures on account of shortages of skilled personnel.

Background

It is the expectation of some analysts that BT would generate a large number of new jobs but these would require very specific and highly trained individuals. Some of the earlier studies have shown that this fact has been perceived by the respondents of the survey from high tech firms who rank quality education, and availability of skilled work force as the factors most crucial for successful commercialisation of BT.

The working biotechnologists at present are not trained in BT but are drawn from related fields like biochemistry, molecular biology, genetics, microbiology, botany, plant pathology, virology, biochemical engineering, fermentation technology, immunology, etc. Understanding the importance of the highly qualified and skilled personnel, in order to attract such skills the BT companies in United States of America have allowed freedom to their scientists in pace and direction of their work, and accorded flexibility of work environment akin to the academic institutions such as universities and research organisations functioning in the public domain.

The human resource requirements evolve sequential to the developments in the field. A rapidly growing field has a different rate of increase in manpower requirement compared to a slowly growing or stabilizing field or already stabilized field area. BT belongs to the category of young and rapidly growing fields. In the same way the nature of the field, such as a theory oriented abstract discipline/field, has less attraction to manpower to work in the field compared to an applied field where the results could be useful to existing or newly created commercial/production activity, increasing the manpower requirements to range from research, teaching/training to production/manufacture, management and marketing. As the stakeholders increase, the demands increase. Thus BT comes under the second category holding a lot of promise for commercial and development activities. Hence, both governments as well as business houses are ready to invest in varied activities related to BT thereby increasing the scope of employment immensely.

The third feature which also determines the manpower requirement and its growth is the field being uni- or multi-disciplinary in nature. This increases the demand of personnel largely to execute various tasks involved in the field (like in BT) at the levels of R & D or production. It requires personnel drawn from various disciplines and having different skills. Thus inherently skill intensive, the personnel-dependent nature of BT makes it a worthy subject for studying the dynamics of human resource requirement to enable better planning.

The human resource needs of the BT industry which is at the phase of growth would change until it reaches maturity. In the initial phase there would be an emphasis on R & D which later would shift to production related capabilities like bioprocess engineering and from scientific personnel to trained technical personnel. There could be a shift in the user industry of BT, such as from the pharmaceuticals industry, which is the major industry (apart from agriculture) applying BT, for their products and might give way to other industrial sectors like energy, mining, industrial chemicals, food processing etc. Over and above, personnel needs in biotechnology of the future would be determined by the type of R & D, production processes used, products produced and the extent to which BT has made in-roads into various industries.

Personnel needs in BT

Personnel needs in BT change with the maturation of the industry. Each stage has its own specific skill requirements based on the accent on the activities during that stage. Generally three major stages could be identified namely viz.

Early stage - Requires research scientists and supporting laboratory technicians.

Commercialisation stage - Accent is on scale up process engineers, cell culture, fermentation specialists, separation and purification specialists, analytical chemists, clinical scientists, regulatory affairs experts and financial analysts.

Full scale production stage - Increased requirement of technicians at various levels, production chemists, chemical engineering, quality control specialists, marketing managers, business specialists, etc.

In keeping with the needs and jobs of the above stages the qualifications of the personnel vary:

First stage - Highly professional PhDs, post doctoral level personnel with the accent on R & D being dominant in the task or job.

Second stage - The emphasis shifts to production and quality control positions and these personnel with qualifications in bioprocess biochemical engineering, analytical chemistry, enzymology, etc.. are needed in addition to R & D jobs. This would also shift the hiring pattern from PhD level researchers to technicians [5]. There would be an increase in the intake of personnel at the master's level and bacheolors' levels in biology and biochemistry. There will be an increase in the opportunity for two-year associates of applied science degrees.[6].

Third Stage The personnel requirement shifts from researchers to managers and marketers. i..e.. personnel who can get products out to markets rather than products out of laboratories [7 & 8] in BT companies.

The other determinant which can influence the changes in the nature and composition of the personnel required by the BT firms is the development of different sectors of BT industry in the given country. The differences could be at two levels namely, (a) specialization/discipline (b) qualification keeping in view the range and concentration of positions.

143

International Scenario

The inherent quality of BT R & D and production has made it necessary that irrespective of the stage of development of BT, the skills have very significant implication as these are constantly required to be updated and could not be mechanised or automated and requires inputs from the high science and theory based knowledge and information to work as well as trouble shoot at various functions like R & D, production and manufacturing. Even at marketing, management and extension functions, which involve coping with new regulatory norms and protection of intellectual property rights these inputs, have a major role.

To cope up with the requirements of skilled manpower in new and emerging areas of biotechnology the attempts made in the USA include taking up new initiatives in biotechnology training. A number of such new initiatives were taken in the 1980's by the American colleges and universities to meet the education and research needs of the upcoming biotech industry. According to a survey by OTA, 49 different US colleges and Universities have taken about 60 such initiatives. The identified initiatives in training in biotechnology include programmes ranging from 2 year applied associate of science degrees to short courses in particular biotechnologies designed for professional scientists.

The other types of initiatives which could be situated outside academic institutions in the research centers of biotechnology (University sponsored). These generally do not sponsor courses but lend their support to biotechnology education and training through access to equipment, faculty development and providing research training/opportunities for both graduates and undergraduates. They also serve as a focal point for discussions as the best means to educate and train new biotechnologists.

Another interesting and important fact relates to the involvement of industries in the development of these programmes. According to the OTA survey of 1988, out of the 41 programmes of above mentioned nature 34 have been evolved in consultation with industry. Consultations with industry included surveying local biotechnology companies and sending programme proposals to industry representatives for comments. But such coordination among the developers of different programmes was not observed. Though the response of the industry to the personnel turning out of such programmes could not be observed, the personnel coming out of these new initiatives faced no difficulties in finding employment in their fields of training.

While some of the strategies adopted by different countries seem to be of consequence for a short term, the American experience seem to hold key for long term strategy to take care of reducing the mismatch in the demand and supply of skills required by BT industry. The strategy involves partnership of the industry and academic institutions with certain amount of input from the research organisations. The nature of involvement could be varying while the benefit is shared. It is also observed from the existing literature that the US has been the most successful country in the commercialisation of BT and has sufficient manpower to achieve this, indicating the validity of partnership between all the three actors in the system involved in the specific skill development required by the BT industry. It can also be noticed from the survey of the efforts by different countries other than the USA that in most of the countries there had been an interaction between the BT industry and University system to abate crisis of shortages in the skills experienced or envisaged by the industry.

Different studies in the past, show that though there is an agreement among different forecasting groups about the fact that there will be an increase in the need for personnel in BT in the nineties, there is no agreement about the types of jobs which might be available and the type of training required for these jobs. OTA (1988) estimates are that the personnel who would be employed in companies involved in BT based product commercialization which includes both dedicated BT companies and large diversified companies, would be about 36,000 of whom slightly more than a half would be scientists and engineers (this is a 5 fold increase over the situation in 1983).

According to a report of a survey of California firms using new BT, by Blumenthal et al (1987) the professional (scientific) and technical personnel occupy 63% of total manpower, while clerical workers (17%), managers (15%) and floor level production and maintenance workers would be the rest of the total BT labour force (5%).

Indian Scenario

Biotechnology is a fast emerging multidisciplinary area of science. Using the rapidly developing modern techniques and tools of biotechnology, processes and products are being announced at regular frequency. For planning, promotion and coordination of BT programmes in the country, the Department of Biotechnology (DBT) was set up in February 1986.

Human resource in Science and Technology (S & T) is an essential input for industrial development of a country. India has been put in such an anomalous position that while there is surplus of S & T personnel qualified in certain disciplines (specially traditional subjects), there is a shortage of trained personnel for areas like biotechnology, energy etc. To take care of the deficiencies in the human resource for Biotechnology, the Department of Biotechnology (DBT) under the Integrated Manpower Development Programme, organised the following programmes in 1986:

- Post-Graduate, Post-Doctoral and Post-M.D./M.S. Programme
- Biotechnology Associateships-National and Overseas
- Short-term training programmes
- Technician training programmes
- Training Programme for industrial R&D personnel
- Post M.Sc. and M.Tech. training in industries
- Visiting Scientists from abroad programme

145

- Biology scholarship and National Bioscience awards
- Biotechnology Publications
- Popular lectures by experts
- Exhibitions
- National Science day
- Production of films
- Seminars/Symposia/Conferences

Major source of data for the following programmes has been Annual Reports of Department of Biotechnology [9] for the years taken for study and primary data collected from various universities, and research institutes which conduct the PG courses and short-term training courses. The data for industrial training was obtained from Biotechnology Consortium of India Limited (BCIL). Earlier studies in this area by Visalakshi & Sharma (10) and Shrivastava et al [11] also were consulted for data and projections made earlier.

Post-graduate and post-doctoral programmes

A model system of Post-Graduate and Post-Doctoral teaching in biotechnology has been evolved in 45 selected leading universities / institutions all over the country in collaboration with the UGC, Ministry of Human Resource Development, ICAR, ICMR and Department of Ocean Development (Table 17.1). These universities / institutions have been selected on the basis of expertise and infrastructure facilities available in the institute, the major ongoing R & D programmes in related areas, grants received from other funding agencies and proximity of other institutes/ national laboratories engaged in biotechnology research. The Department has played a catalytical role by establishing these model institutes and this has stimulated several other institutes/universities to establish such institutional facilities on their own. Top grade students with background in basic and applied biology, chemistry, physics etc. are selected on all India basis and all selected students are given studentship. Krishnaswamy committee on Integrated Manpower Planning, which submitted its report in 1988, has recommended continuation of the programmes and also introduction of some new institutions conducting post-graduate programmes in biotechnology.

The postgraduate programmes are continuously being monitored by the Department through meetings of the Advisory Committees, meetings of the Course Coordinators and the National Board of Studies. A lot of infrastructure by way of equipment and laboratory facilities have been built up in these institutes. The general standard of teaching as well as research has improved in these universities and in the country as a whole.

Table 17.1 List of the Universities Offering Post-graduate Courses in Biotechnology Sponsored by DBT

1985-87
1. Madurai Kamaraj University
2. Maharaja Sayaj Rao University
3. Pune University
4. Jawaharlal Nehru University
5. Jadavpur University
6. Benaras Hindu University
7. IIT-Kanpur
8. IIT-Delhi
9. IIT-Bombay
10. AIIMS-Delhi
11. IVRI, India

1987-1988
Nil

1988-1990
12. GB Plant University of Agricultural and Technology
13. Tamilnadu Agricultural University
14. Goa University
15. Assam Agricultural University

1990-1992
16. AM University
17. Anna University
18. Bose Institute
19. Roorkee University
20. University of Hyderabad
21. Gurunanak Dev University
22. DAVV

1992-93
Nil

1993-94
23. University of Bombay

1994-1998
24. Tezpur University
25. S.V. University
26. Birsa Agricultural University
27. Gujarat University
28. Himachal Pradesh University
29. Jammu University
30. Calicut University
31. Gulbarga University
32. Mysore University
33. UDCT
34. Punjab University
35. Thapar Institute of Technology
36. Pondicherry University
37. Banasthali Vidya Peeth
38. TNV AS University

1999-2000
39. Allahabad University
40. Himachal Pradesh Krishi Vishwavidyalaya

2000-2001
41. Guru Jambeshwar University
42. Kashmir University
43. Kumaun University
44. Indira Gandhi Agricultural University
45. Marathwada Agricultural University

The courses the department is supporting at M.Sc./M. Tech. level in the selected universities/institutions are:

- 23 M.Sc. courses in general Biotechnology.
- 6 M.Sc. courses in Agricultural Biotechnology.
- 1 M.Sc. course in Medical Biotechnology.
- 1 M.Sc. course in Marine Biotechnology.
- 4 M.Tech./M.Sc.(Tech.) in Biochemical Engineering, Bioprocess technology and Biotechnology.
- 3 Post MD/MS course in Medical Biotechnology
- 2 One year PG Diploma in Molecular & Biochemical Technology and Clinical Biochemistry and Biotechnology

The Table 17.2 gives the number of admissions made each year to the PG programmes in biotechnology from the year 1985-86. Based on data available the growth was quantified using modelling approaches.

Table 17.2 The number of admissions to the PG courses in BT (1985 to 2001)

Year	No. of admissions
1985-86	64
1986-87	111
1987-88	147
1988-89	179
1989-90	134
1990-91	154
1991-92	192
1992-93	196
1993-94	300
1994-95	350
1995-96	350
1996-97	400
1997-98	450
1998-99	450
1999-00	500
2000-01	550
Total	**4,527**

Estimation regarding the supply of manpower through the PG courses has been done by using the trend extrapolation method using the statistical package Minitab* The data on the out-turn of manpower for the years 1985-2001 for the PG courses in Biotechnology has been obtained from the various Universities and DBT.

Post M.Sc. and M. Tech. Training in industries

It has been found that many industries insist on some industry working experience for providing employment to students after postgraduation. It was therefore proposed to have a programme for training courses in biotechnology sponsored by DBT which would be helpful to them in finding employment. Some may even get absorbed in the same industry. It may also help some trainees to start industries on their own.

Students passing out of the post-graduate courses in BT are selected and placed for training in various biotech based industries for a period of six months. This scheme is being implemented through the Biotech Consortium India Ltd. (BCIL), New Delhi. Initially in a year around 25 students were selected for training at various (Table 17.3) industries.

Table 17.3 Number of post M.Sc and M.Tech students placed in industrial training (1993 - 2002)

Year	No. of students placed for training
1993-94	18
1994-95	30
1995-96	20
1996-97	24
1997-98	20
1998-99	25
1999-00	22
2000-01	30
2001-02	45
Total	**234**

From the above table, it is evident that in the period from 1993-2002 an average of 27 persons have been placed in the industry for training. The highest number being in 2001-2002 with 45 students having got the training. Compared to the number of students who pass out of post graduation, this number who get selected for training is very small.

Technician training programmes

Well trained technicians are an asset to any good research laboratory. They play a vital role in research laboratories by making arrangements for/and performing standard/routine experiments as desired by senior scientists, project/programme leaders and principal investigators. This enables the scientists to devote their time more effectively for productive research. There is a dearth of technicians properly trained in modern laboratory techniques.

The Department of Biotechnology has formulated two types of Technician training schemes viz., (i) Short-term (3 months duration) and (ii) Long-term (1 to 2 years duration). Under the Short-term scheme Technicians/Technical Assistants/Technologists in research laboratories are provided training for a duration of 2 to 3 months in order to upgrade their skills.

Two long-term Technician Diploma Programmes each of one year's duration have been started with the financial support of DBT at Madurai Kamaraj University, Madurai and Sri Venkateswara College, University of Delhi, New Delhi. The former course started during 1989-90 and the latter started in the academic session of 1990-91. Number of admissions in these courses during 1989-90 to 1998-99 is available in Table 17.4.

Table 17.4 Number of admissions for technician training programme 1990-1998

Year	No.of admissions
1990-91	21
1991-92	20
1992-93	22
1993-94	21
1994-95	24
1995-96	22
1996-97	20
1997-98	27
Total	177

Short-term training programmes

The Department sponsors a few short-term training courses of 2-4 weeks duration annually to impart practical training on newly emerging research techniques in areas of modern biotechnology. The objective of the programme is to provide practical hands-on training to mid-career scientists for meeting the requirements of newer R & D techniques in Biotechnology within the country. The Department identifies areas of training as well as universities, national laboratories and other research institutions which have the desired level of expertise and adequate infrastructure for organising such courses. Mid-career scientists/technologists engaged in research activities sponsored by their parent institutions are given preference for attending the courses with the idea that the participants on returning to their parent institutions could utilize the techniques/knowledge gained during the courses in their R & D and teaching programmes.

Funds are provided for purchase of laboratory equipment/accessories spares, consumables and other recurring expenses needed for organising the courses. Provision has been made for inviting 3-4 guest faculty from within the country and if necessary one foreign expert also, in identified gap areas for each of the courses, so that the participants in each course get an opportunity to interact with the leading scientists working within as well as outside the country in particular research areas. From the year 1986-1998 about 70 institutions together have conducted about 185 short-term training programmes sponsored by DBT. (Table 17.5)

Table 17.5 Number of short-term training programmes organised during 1986-2000

S.No	Year	No. of Programmes
1	1986	9
2	1987	12
3	1988	11
4	1989	14
5	1990	14
6	1991	16
7	1992	11
8	1993	14
9	1994	12
10	1995	13
11	1996	13
12	1997	9
13	1998	14
14	1999	11
15	2000	8
	Total	181

Biotechnology Associateships

This scheme was started with a view to build a strong research base in identified priority areas of biotechnology. It is hoped that the personnel trained under the programme would not only strengthen the research programmes of their institutions, but also train their co-scientists. It is expected that the programme would be helpful in filling the gap in internal competence in this area and in meeting the requirement of highly skilled and trained manpower for the ongoing biotechnology R & D activities in the country.

Two kinds of associateships, namely, National Associateship and Overseas Associateship are awarded every year.

National Associateship

The national associateship is awarded to work in leading Indian research institutions other than the candidate's parent institution. The scheme also has a provision for training in an overseas laboratory for a period of 3 months provided one completes 9 months of associateship in India successfully and the overseas component of the training is justified. The associates join premier institutions in India like NII, CCMB, IISc., IARI, MKU etc. to work under the programme. The associates have worked in frontier areas like development of transgenic plants and animals, immunodiagnostics, plant tissue culture, pollution control etc. The associates are selected from various Universities and research institutions of ICAR, ICMR, CSIR, etc. Details of number of scientists who were awarded these associateships is available in Table 17.6.

Table 17.6 The number of national and overseas associateships from 1986-2001

Year	No. of Associateships Awarded	
	National	Overseas
1986-87	-	15
1987-88	24	17
1988-89	25	20
1989-90	21	25
1990-91	24	27
1991-92	20	25
1992-93	20	20
1993-94	20	25
1994-95	25	23
1995-96	20	26
1996-97	23	25
1997-98	26	26
1998-99	20	25
1999-00	15	20
2000-01	13	22
Total	**296**	**341**

Overseas Associateship

The Overseas associateship is awarded for pursuing research in frontier areas in reputed institutions overseas. It also provides an opportunity to the scientists to interact closely with the leading scientists of repute and get useful exposure to the recent developments that are taking place in the field of biotechnology. It helps the scientists to attain experience to meet the global standards. These awards are made under two categories, namely, long term and short term. The long-term associateship is tenable for one year, while the short-term associateship is awarded for 3-6 months only. Applications are invited through all India press advertisements and scientific agencies viz. ICAR, ICMR, CSIR, UGC, DST etc. Applications are short-listed by a screening committee and selection committee. Table 17.6 has the details of overseas fellowships.

Analysis and Trend Extrapolation

Based on the past trend projections for three initiatives which involve new talents viz., introduction of PG courses, Technician training course and industrial training were done using the Statistical package Minitab*. The model fit was different in the three cases in the case of PG courses quadratic model had the best fit, while in the case of technician training and for the industrial training linear model had the best fit. The projections made for the year 2010 under these three types of efforts showed that

- No. of students who would be admitted in PG courses in Biotechnology would be around 1100.
- No. of students who would be benefitted by the technician training would be around 30-32.
- No. of placements of students in the industry would be around 160.

The underlying assumption for these predictions have been that the rates of change in intake would continue to be same. In the case of PG courses, addition of institutions will be at the same rates as in the past and also the earlier admissions would have an effect on the current year admissions./placements.

Discussion

Through the above programmes we see that the three major approaches to capacity building has been addressed, namely

- Developing new talents in Biotechnology
- Reorientation of exiting scientific human resource
- Upgradation of the existing human resource in the area of biotechnology.

It could be seen that while PG courses in BT, technician training programmes and Post MSc, M Tech Industrial training aim at the first category, Post-doc programmes, short-term training programmes come under the second and BT associateships promote upgradation of existing BT human resource manpower developing new talents in Biotechnology.

153

The Post Graduate courses in BT are very important for laying the foundation for the required manpower in BT, the most important in this connection is the number of Universities which offer this programme (Table 17.1) and the number of seats they offer per year. It is seen that the number of Universities has increased from 11 in 1987 to 15 in 1990 and 22 in 1992 and to 39 in 1998 and reached 45 by 2001. The number of students admitted each year has followed an quadratic model as confirmed by the trend extrapolation model. The trend extrapolation indicates the need for either increasing the number of Universities, facilities and the very influencing interaction between the students passing out of these courses, in order to maintain the trend. From Fig. 15.1 and the table attached to it we see that 1100 is the number of admissions expected in 2010 if the trend continues to be same.

The training in industries for the Post M.Sc and M.Tech students is very important for aiding them to use their knowledge in their employment or to even start an industry enterprise of their own. As compared to the current pattern of around 500 students annually the number of students selected for training in various BT related industries is very small. On an average only 27 students have received the training in the past 9 years and out of them only a few have secured placement. So it indicates the need to increase the opportunities for industrial training. More number of industries may have to come forward to take students for training students for the benefit of the country.

In the present day, technicians are preferred for employment by the industries. From that angle when we look into, it is important to train more number of technicians. But unfortunately only two institutions offer this type of training programme and the out-turn of technicians is only 22 per year on an average. This indicates the necessity of giving more attention to expand this training in more number of institutions. The trend extrapolation shows that it is linear and hence even over the next 10 years hardly the number increases by 10 which indicates serious and urgent attention to increase the facilities for training technicians.

Reorientation of the existing scientific manpower

On an average 13 different programmes are conducted each year with participation of 13-15 students per programme. The maximum number of programmes has been in the field of genetic engineering followed by programmes in biochemical engineering./enzyme engineering, followed by hybridoma technology and plant tissue culture. The field of Genetics is proving itself very efficient in the development of transgenic plants, and the rDNA technology is playing a major role in engineering enzymes etc. So it is quite reasonable that more number of programmes are organised in this field to help the scientists to enhance their professional capabilities (Table 17.7).

Table 17.7 Training programmes conducted from the year 86-98 in different areas of Biotechnology

FIELDS	1986	1987	1988	1989	1990	1991	1992	1993	1994	1995	1996	1997	1998
Genetics/Engineering	2	3	3	5	7	2	3	4	6	5	3	5	1
Hybridoma Tech./Immunodiagnostics/Immunology	0	2	0	4	2	3	3	2	3	0	6	1	5
Plant Tissue Culture/Plant Molecular Biology	2	2	3	2	2	4	4	4	1	3	1	1	1
Biochemical Engineering/Bioprocess Technology/Enzyme engg.	4	4	4	3	3	6	1	4	1	3	1	2	4
Others	0	0	1	0	0	1	0	0	0	2	2	0	2

It is observed that maximum participation is from the Universities, followed by the research institutes and finally very meagre participation from the industries (See Table 17.8). The participants from the Universities is six times that of industries and other private concerns, and participation of research institutes is about four times that of industries. The benefits of good research in the form of a good product can be only made through the industries. So the industry must be encouraged to increase their participation. This could be accomplished by involving the industries in DBT efforts, giving incentives for nominations to training, etc.

Table 17.8 Affliation of Participants in the Short-term Training Programmes in Biotechnology

Year	Affiliation		
	Universities	Research Institutes	Industry
1986	46	33	13
1987	67	50	3
1988	37	13	7
1989	27	22	18
1990	80	75	4
1991	96	64	20
1992	55	41	5
1993	54	39	12
1994	56	26	6
1995	72	35	11
1996	46	32	10
1997	67	29	11
1998	91	60	11

Upgradation of the present BT manpower

The Biotechnology associateships as seen are very important for competing at global levels. This programme plays an important role in encouraging the scientists to upgrade their skills. The average number of National Associateships has been 21 per year and average of overseas associateships has been 23 per year, over the years 1986-1999. There is a need to increase the number of associateships so that more number of scientists benefit the country by qualifying themselves to compete internationally.

Conclusions

The major points which emerge out of this present study based on the survey methodology employed and discussed are :

- The present trend in the manpower coming out of the PG courses organised in various Universities by DBT is following a quadratic model due to the increase in the number of institutions offering this programme over the last 10 years.

- The Post M.Sc. and M.Tech. industrial training is provided only to a handful of students compared to the number of students who pass out every year. Even out of the students who received training only a few got placement.

- Similarly, it is seen that the facilities for training more number of technicians is not available. Facilities are available only to train an average of 20 technicians per year.

- The participation of industries in the short-term training programmes is very low compared to that from universities and research institutes.

- The number of BT associateships being provided now gives opportunities for about 20 scientists each per year to work in National laboratories and overseas.

Suggestions to increase production of human resource with the requisite qualification are :

- Universities could be informed by DBT of the possible shortages and initiate or strengthen courses which would be able to produce personnel in adequate numbers with required training competence.

- The BT related industries / enterprises should be encouraged to get benefitted by Mid career training and retraining through short and mid-term training courses. The industries should have a say in the designing of the training courses and also cosponsor them thereby having an opportunity to get the manpower with desired skills.

- The DBT should pay very serious attention to increase facilities to train more technicians which is the need of the hour.

- To get the Indian biotechnologists working abroad and trained in recent advances in BT may be offered incentives to come back and invest and work in the home country.

- To facilitate efforts to increase rates of absorption of trained manpower for the jobs for which they are trained.

- Increase jobs at the level of training like initiate courses in many colleges for postgraduate diploma courses. This effort while increasing the number of personnel of required skills/qualification available for the industries would also increase the number of jobs available for PG's in courses related to BT for the purpose of training.

Acknowledgements

The authors acknowledge with thanks the efforts of Ms. N. Arthi, a BITS trainee during 1999 who collected data for some years under different programmes and Prof. P. R.Lakhmikanthan of Indian Statistical Institute for kindly helping with projection for 2010 using Mintab and interpretation of the analysis.

References

1. Biotechnology companies gear up for marketing of products with new executive talent (1985) **Applied Genetics News,** Vol 5, no. 8, PP 10-11.

2. Biotechnology firms reshape staffs for move into marketplace, (1984): **Chemical and Engineering News,** Vol 62, no. 25, pp 10-11

3. California State Senate Office of Research Silicon Valley II a (1985) Review of State Biotechnology Development Incentives, Sacramento. C. A. Joint publication office.

4. Department of Biotechnology, Govt. of India, New Delhi, Annual Reports 1985-86 to 2000-01.

5. Gelhart F. and Cass M (1985) How to select a site for your biotech plant, **Genetic Engineering News.** Vol 5, No. 2, p 13

6. Industrial Biotechnology Association, Radford Associates (1987) Biotechnology compensation and Benefits Survey Seminar, Washington DC.

7. Klausner A and Fox J (1986) Some bird's-eye views of agbiotech-88. **Bio-Technology,** Vol 6, No,3, p 243-244

8. Silvestri G R and Lukasiewicz J M (1987) Projections 2000; A look at occupational employment trends to the year 2000, **Monthly Labor Review,** Vol 110, No.9, pp 46-63

9. Annual Reports of Department of Biotechnology, Govt. of India.

10. Visalakshi.S and Lalita Sharma "Manpower Projections in BT in India by 2000 A.D: Final Report based on Phase II survey", July 1995

11. Shrivastava M.P et al, "Development and Planning of Biotechnology Manpower in India", Agricole Publishing Academy for IAMR, 1994.

18

Human Resource Requirements in the Biotech Sector
An Industry Perspective

M. A. Mukund
AstraZeneca, -India.
No. 277, T. Chowdaiah Road, Malleswaram, Bangalore 560 003
mandyam.mukund@astrazeneka.com

Biotechnology, the science to use and manipulate natural living systems to derive products or information for commercial applications, has attracted a lot of attention in India in recent times. After the Information Technology revolution in the nineties, Biotechnology has been projected as the next big wave for business in the new millennium. Many states like Karnataka, Andhra Pradesh, Maharashtra and Tamilnadu have taken the lead towards establishing a base for biotechnology to attract global investments in their respective states. The formation of "Biotech parks" is one such initiative in these states.

Biotechnology made its transition from academic laboratories to corporate board rooms in the west atleast 15 to 20 years ago. Recombinant DNA and genetic engineering have fueled the explosion of commercial potential in biology; in fact the central dogma of molecular genetics can, in a lighter vein be read as : *DNA makes RNA that makes protein which makes money*. However, it is important to remember that the science of biotechnology is deeply rooted and founded on basic principles and concepts of biology; ideas arising out of exploratory research have been translated to applications to bring in revenues and profits. Managing a biotechnology business is similar to managing other businesses, but one factor that assumes relatively more significance in biotechnology is people. Often, a start-up biotechnology enterprise revolves around a basic scientific idea, which arises through scientific thinking, either individually or collectively. It is this core need of idea creation which is at the heart of the biotechnology business - a pipeline of new scientific ideas has to constantly feed into the business, where those that have the maximum commercial potential have to be identified. People have to translate ideas into action.

At the core of a biotechnology enterprise, its human resources are key to its success. In India, they are sourced from universities, educational and research institutions either within country or from overseas. At junior levels, typically the scientific resources are post-graduates (M.Sc, M.Tech, M.Pharm etc.) in microbiology, biochemistry, cell biology etc. from various Indian universities and educational institutions. At mid and senior levels, organizations look for wide-ranging and in-depth scientific experience, who have specialized in specific areas of relevance, within the above domains, and are willing to search for suitable talent from across the globe. Scouting for such talent involves time, effort and money. The domains are in most cases overlapping and hence their expertise becomes inter-disciplinary. In a very broad sense, these have been knowingly or unknowingly differentiated as the doers and the thinkers / planners. An unwritten philosophy which is based on general management theories seems to be that the junior scientific pool comprises of "soldiers" at the front-line or site of action, to execute scientific experiments, while senior and relatively more experienced scientists act as the "strategists" in terms of thinking, planning and conceptualizing. While this could be true and justified in most situations, organizations that wish to maximize possibilities and widen the net for capturing ideas would do well to harness the talent that is available at all levels to unleash creativity and innovation from every single resource. Unfortunately, our education system has done precious little to address these vital issues which are paramount for success in any industry today, let alone biotechnology.

Biotechnology as a subject in modern education has found a place in almost all Indian universities and educational institutions. In some States, biotechnology as a course has been introduced at the undergraduate level, and sometimes projected as a academic programme to train young entrepreneurs. This is quite out of tune with the need of the hour. What is essential is to build the foundations of basic modern biology in young undergraduates and spur the interest in them to pursue specific disciplines within biology (such as microbiology, cell biology and biochemistry) in post-graduation and further on in doctoral studies. Biotechnology courses in India as they exist today offer an abbreviated amalgamation of various disciplines, often without doing justice to these to the extent required. These disciplines are given superficial coverage in the courses offered, as a result of which students are unable to gain an in depth understanding of any of the underlying streams in biology. In such a situation, it is difficult for a post-graduate student to develop a wholesome and integrated perspective in the disciplines, whereby basic fundamentals may not be grasped entirely. At the school level, biology is taught by some state-level educational authorities in a fragmented manner, which spills over to college or undergraduate levels. Excessive emphasis on classical biology in some of these systems lead to descriptive and non-analytical learning modes, where information is disseminated as a string of facts to be memorized and not as a science that requires an analytical approach and application of the human mind. A paradigm

160

shift in our educational system is required whereby biology be treated as an exploratory, experimental and analytical subjcct as much as physical sciences or mathematics. It is important to differentiate between classical biology, analytical or quantitative biology, since biotechnology's core focus is on the latter framework. This is not to decry the importance of classical biology, which should certainly form the initial foundation on which analytical biology should be built.

From the point of view of the biotechnology industry, new recruits at entry-level positions (typically M.Sc, M.Phil or M.Tech in any branch of biosciences) coming out of universities and institutions need to "unlearn" quite a bit before they are reoriented some of the fundamentals in the subject. Students need to understand that in modern biology, "facts" as presented in textbooks are not static-these represent the current thinking, which may change in the light of new scientific information that may come in at a later date. Budding biotechnologists with a grounding in mathematics at the undergraduate and / or post-graduate level have a distinct competitive advantage over others. Emphasis on quantitative and statistical methods along with their applications is imperative.

Practical, analytical and problem-solving abilities in basic biochemistry, microbiology, cell biology and molecular biology are absolute essentials; educational courses in these areas should be oriented towards experimental approaches to solutions, case histories and hypothetical challenges to stretch the human mind. Training is often imparted on the job in performing experimental protocols, data interpretation and drawing scientific inferences and conclusions. More attention is often paid to gaining hands-on expertise in techniques and procedures; one must realize that these can be learnt along the way. In contrast, analysis, interpretation and making valid scientific judgments based on precise observations and data are more difficult to cultivate, and need constant practice and perseverance. A student of science needs to develop the ability to link data and conclusions of one experiment or set of experiments to those derived from another in a meaningful manner and propose concepts or hypotheses. Wherever possible, he should be able to establish a logical and sequential scientific chain of thought in his work. In the long run, it is this skill along with many others which will identify him as a scientist. Towards this objective, students need to be encouraged to take up experimental projects for short periods of time in industrial or established academic research laboratories. Here is where the industry can make a deep impact in collaboration with universities. By providing opportunities to students coming out of universities to undertake short-term scientific projects, for example, as summer training workshop in their laboratories, industry can facilitate their experimental learning, which has maximum impact in terms of effectiveness. As a corollary to this, universities should make it mandatory for students to gain such project-oriented short-term experience in an industrial research or an established reputed academic setting. Such programs can fulfill two objectives; firstly, a social responsibility by the industry and secondly, promotion of industry-academic interaction and collaboration.

In a biotechnology industry, scientists need to focus towards differentiation between generation of data and generation of meaningful data that advance projects. In this context, project management and advancement skills become critical. Often these crucial issues are overlooked by some companies, where it is taken for granted that experienced scientists can manage projects. Project management skills span a wide range of issues such as goal-setting, work-planning, assessment of resource and infrastructure requirements, time estimation, milestone identification, people management, communication and finally decision-making abilities, specifically "stop-go" decisions. While it may be heart wrenching to close projects which have originated from original ideas due to any business reason, scientists must learn to come to terms with such decisions and channel their energies towards new business imperatives. This is easier said than done, since scientific ideas arise from deep-rooted thoughts in an individuals scientific minds. Pet theories overshadow the individual scientific ego, which are quite difficult to resolve.

Scientists are products of academia, often with relatively less commercial background or orientation. Since a biotechnology business has to be operated like any other business, scientists need to develop commercial focus with objectivity towards the business. In this context, scientists would do well to develop a reasonable degree of financial discipline with reference to their research, specifically in terms of awareness and working within budgets. Cost control issues in science can pose dangers of conflict in a scientific environment. Striving for a balance between controlling costs and pursuing scientific goals will remain a challenge for the biotechnology industry, since this industry is capital, revenue and resource intensive.

Those who are recruited for mid-level positions in any company form the core of the organization, on which deliveries towards the organizational goals depend on. They need to equip themselves to design experimental approaches, envision scientific start and end-points, seek out critical information and link data and inferences from different overlapping disciplines. Deriving and proposing new ideas and the way forward, along with time management are critical abilities that are needed to advance scientific projects.

At a more fundamental level, individuals who wish to pursue their careers in biotechnology must arm themselves with theoretical and practical knowledge on biosafety and radiation safety issues, environmental impact of laboratory-work, patents and intellectual property issues as well as literature survey methods. The importance of documentation must be stressed at all levels. Clarity in this vital issue, especially in terms of maintaining scientific records and reports is paramount. Scientific documentation must reflect the twin facets of reproducibility and reconstructability of data. Our educational system in biotechnology must address these topics. Today, the information technology revolution has opened up new frontiers such as bioinformatics and genomics; it must be remembered that these are tools available for accelerating research in modern biology and not endpoints by themselves. All said and done, problems in biology are solved on the laboratory bench, which are supported and to some extent facilitated by computer-aided studies.

Unlike the information technology industry, where the gestation-time of a project is relatively short, and turnover of the human resources is high, ideally, the association between a biotechnology company and a scientist can be expected to be a long one. A scientist grows to become a knowledge-asset to the organization over time, and it is crucial for organizations to retain them. Losing creative talent can prove to be costly, especially in terms of replacement and the loss of knowledge and expertise that may have been built up over the years. Continuity in the progress of a project could be a casualty in such times. Biotechnology is an expensive business to run, but then, the returns are high when they occur. Hence, industry would do well to deal with scientists as strategic long-term partners towards fulfilling a common vision and mission.

Alternative career paths which are important to a biotechnology business should be major considerations. Some of these are careers in business development management, regulatory affairs, intellectual property, human resource management as well as public and media relations. Although science would be at the core of the biotechnology industry, these functions require competent individuals who understand the biotechnology business as much as a biotechnologist. Our education systems need to build and develop skills in these areas for aspiring students and professionals who may either not be suitable to the scientific stream or who may seek to choose these alternative careers. Marketing biotechnology-derived products or services requires competent professionals who may preferably originate from the scientific stream. Their customers in most cases would be scientists or technical professionals; technical marketing would be needed here. A biotechnology company would also need professionals who would possess expertise in regulatory frameworks and intellectual property, with specific knowledge of the legal environment and context in which the company would operate. Considering the global attention that biotechnology has attracted in recent times, such a company must have a comprehensive strategy to effectively manage public relations, and especially the media. The image of the industry hinges on this crucial function to a considerable extent. Finally, human resources management needs to be integrated into the business strategy of the company since people matter most in this endeavour.

Organizations would do well to invest in people development, especially for their scientific pool. People development is one of the principal elements of motivation. Assessment of specific training needs, coupled with medium to long-term development plans will add value to a scientific career. Performance management systems must be designed to take this issue into account. Development in contrast to training is more abstract. Training focuses on procedures and protocols, whereas development is a continuous on-going process without a finite end, but with a perspective to meet future challenges. It can be viewed as a confidence building measure which continuously adds to all round capabilities of an individual. Development plans for junior scientists can address ways to help them face complex scientific issues and problems with a long-term aim to become experts in their fields. They have to graduate over time from merely conducting experiments to hypothesizing scientific concepts and designing experimental approaches to test the validiity of such concepts. Development plans for experienced scientists can enrich and diversify their experience to broaden their scientific horizons as well as build their leadership skills. At senior levels, as in any other industry, management and leadership skills are most important with objective decision-making abilities as the key issue. Scientific though, leadership towards novel and pioneering directions, in specific disciplines.should be another major thrust area for development of senior scientists. A word of caution here is to maintain the focus on the commercial objectives of the biotechnology company while embarking on development plans to ensure balanced alignment between the two priorities.

Apart from technical and scientific skills that a scientist in a biotechnology company must possess, soft-skills and competencies are two other major areas that need a lot of ttention. These areas are sometimes neglected by some organizations or trivialized. Communication, interpersonal skills and ability to work as part of a team are absolute essentials. These factors ultimately depend on the personality and conscience of the individual; training and development programs in these areas can only lead the horse to the water, but cannot make it drink. While individual contributions are of course important, since ideas would originate from individual minds, successful organizations harness the collective creativity and innovation potential of their teams by encouraging scientists to pool their minds together to work towards unifying goals. The second area that scientists must focus on is that of competency development. These are behavioral indicators that are essential for successful performance at work. Competencies can be classified into clusters, whereby those essential for different functions and levels can be defined at the outset, for example: self-managing competencies, thinking competencies, achieving competencies and people-management competencies.

The biotechnology industry is prone to rapid change and advances. Scientists in this field need to be constantly exposed to new technologies and information. An in-house physical and ctronic library, equipped with relevant books and scientific journals are basic requirements for any organization. This apart, scientists need to participate in national and international seminars, workshops and symposia to exchange information, communicate with peers and to gauge trends in the field. Sabbaticals and secondments, whereby scientists are facilitated to work for defined periods in laboratories other than their own on different or related areas also add value to both the individual and the organization. The organization can benefit by applying the new tangible knowledge that the scientist may have gained through such processes. For the individual, his professional status and knowledge are enhanced by participating in such a program. However, the organization and the individual must define the. objectives of such a program beforehand and ensure alignment and commitment to mutual goals, since high costs and commitments by various stakeholders are involved. These stakeholders, who are investing their time, effort and money are the sponsoring organization, the host institution and the individual who is to participate in the sabbatical or secondment.

It is important to educate scientists entering the field on scientific ethics and philosophy, along with an overview of the scientific process. Values that an ideal scientist will imbibe are confidentiality, scientific integrity towards data-generation, documentation, discretion in information handling and above all trust and respect for peers and colleagues. The individual must realize that the organization places a significant amount of confidence and trust while assigning responsibilities for carrying out his/her scientific functions. Likewise, the individual has to trust the organization that his/her legitimate interests and scientific needs will be protected or provided for. Although conflict and debate are integral to the scientific ethos, these have to be managed with care to avoid dysfunctionality. The spectre of credit looms large in any successful scientific endeavour and so are the rewards that go along with such success. These seemingly "minor" issues tend to get marginalised in organizational to their detriment. Clear-cut and transparent policies on these management issues for a scientific environment will go a long way in contributing to a healthy atmosphere in the organization.

Finally, with reference to Abraham Maslow's hierarchy of needs theory, organizations that wish to gain the maximum out of their scientific talent must explore ways and means to keep the scientists at the highest levels of motivation possible and aim to help them realize their potential i.e. self-actualization, but within the sphere of organizational goals. The company must also ensure that its "hygiene" and motivation factors are functional and effective. Efficient support systems in terms of responsive, responsible and professional administrative functions facilitate the progress of a biotechnology company towards its stated goals. In this regard, it is important for scientists to recognize and respect the identity and distinctiveness of professional areas other than their own. Often they may be tempted to digress from their own scientific functions into these areas leading to unproductive conflicts.

In summary, managing and motivating scientific human resources are critical factors in the success of any biotechnology enterprise. Our education system must focus on overhauling its programs in biotechnology towards analytical learning, experimental problem solving, data interpretation and scientific visioning. The Indian industry would need to explore ways to attract and retain the best available scientific talent from the global arena. Recruitment processes should be designed to assess the scientific worth of candidates and their relative suitability to organizational objectives and culture, as well as be able to differentiate between awareness, proficiency and in-depth expertise in any particular discipline. Compensation packages and reward systems must seek to attract the best talent, and should not revolve around monetary aspects alone. Projecting a supportive and caring image, work-life balance and scientific excellence are factors to be considered in this context. Training and development programs must address scientific needs at different levels, with the objective of developing cross-functional, domain,-specific and lateral expertise, both in scientific and managerial domains. In addition, for a biotechnology company to flourish, it has to create an internal ambiance and an external image to attract and stimulate creativity and innovation, as well as unleash the scientific spirit to achieve organizational goals. A healthy, supportive, intellectually-rich and participative culture in the company, where symbols of hierarchy are either absent or as subdued as possible, will provide a conducive environment for scientific enrichment and excellence. It is imperative that the Indian biotechnology industry keep pace with global benchmarks in the field, keeping the above parameters in view, if it has to succeed in this newly emerging arena.

WORKSHOP PARTICIPANTS

Inaugural Address

His Excellency Dr C Rangarajan

Governor of Andhra Pradesh
Raj Bhavan,
Hyderabad-500 049

Keynote Address

Dr G S Khush

Principal Plant Breeder & Head

Division of Plant Breeding, Genetics
& Biochemistry,
International Rice Research Institute,
MCPO Box 3127, Makati 1271,
Philippines.

Paper Presentations

Mr Arvind Kapur

Managing Director

Nunhems Seeds Pvt. Ltd.,
Dhumaspur Road, Badshahpur,
Gurgaon-122 001

Dr Deepak Pental

Professor & Director

Centre for Genetic Manipulation
of Crop Plants,
University of Delhi South Campus,
Benito Juarez Road,
New Delhi-110 021

Dr P S Janaki Krishna

Subject Expert (Biotechnology)

Biotechnology Unit,
Institute of Public Enterprise,
O U Campus, Hyderabad-500 007

Dr Julius T Mugwagwa

Programme Officer

Biotechnology Trust of Zimbabwe

Dr M S Lakshmikumaran

Bioresources and Biotechnology division,
Tata Energy Research Institute,
Darbari Seth Block,
Habitat Place, Lodhi Road,
New Delhi – 110 003

Dr T Mohapatra

National Research Centre on
Plant Biotechnology,
Indian Agricultural Research Institute,
New Delhi – 110 012

Mr M A Mukund

Manager – Human Resources

Astra Zeneca India,
No.277, T.Chowdiah Road,
Malleswaram,
Bangalore-560 003

Dr Kahiu Ngugi

Kenya Agricultural Research Institute,
National Dryland Farming Research
Centre,
P O Box 340,
Machakos, Kenya.

Shri K Nimmaiah

Executive Director

P E A C E,
Near SLNS Degree College,
Bhongir-508 116

Mrs Orseline Carelse

The Biotechnology Research Institute,
Scientific and Industrial Research and
Development Centre,
P O Box 6640,
Harare, Zimbabwe.

Dr G Pakki Reddy

Coordinator, Biotechnology Unit

Institute of Public Enterprise,
O U Campus,
Hyderabad-500 007

Dr Prabuddha Ganguli

Vision IPR,
103-B, Senate, Lokhandwala Township,
Akurli Road,
Mumbai - 401 001

Dr Rakesh Tuli

Centre for Plant Molecular Biology

National Botanical Research Institute
Rana Pratap Marg, P B No.436
Lucknow – 226 001

Prof Ram Rajasekharan

Department of Biochemistry,
Indian Institute of Science,
Bangalore-560 012

Dr V Ramesh Bhat

Deputy Director (Sr Grade)

National Institute of Nutrition,
Jamai Osmania,
Hyderabad-500 007

Dr T V Ramanaiah

Scientist, Department of Biotechnology

Ministry of Science & Technology,
Block 2, 7th Floor,
CGO Complex, Lodi road,
New Delhi-110 003

Dr K V Rao

Associate Professor

Centre for Plant Molecular Biology
Osmania University,
Hyderabad-500 007

Dr A R Reddy

Professor

School of Plant Sciences
University of Hyderabad,
Central University PO,
Hyderabad-500 046

Shri V Satya Bhupal Reddy

Executive Director

R E E D S,
17-1-386/S/22,
S N Reddy Nagar, N S Road,
Champapet (PO), Hyderabad-500 060

Dr N Seetharama

Director

National Research Centre for Sorghum,
Rajendragar,
Hyderabad-500 030

Dr M Sujatha

Senior Scientist

Directorate of Oilseeds Research,
Rajendranagar,
Hyderabad-500 030

Dr M Udaya Kumar

Head, Department of Crop Physiology

University of Agricultural Sciences
Bangalore-560 065

Dr Usha Zehr

Joint Director (Research)

Maharashtra Hybrid Seed Company,
Post Box No.76,
Jalna-431 203

Dr B Venkateswarlu

Principal Scientist

Central Research Institute for
Dryland Agriculture,
Santhoshnagar,
Hyderabad-500 059

Dr (Mrs) Vidya Gupta

Scientist

Plant Molecular Biology Unit
National Chemical Laboratory,
Pune-411 008

Dr S Visalakshi

Scientist

National Institute of Science
Technology in Development Studies,
Dr K S Krishnan Marg,
New Delhi-110 012

Public Debate - Discussants

Dr C R Bhatia

Former Secretary, DBT

17, Rohini, Plot No.29-30,
Sector 9-A, Vashi,
MUMBAI-400 703

Dr M V Rao

Chairman, APNL Biotechnology Programme Committee

Biotechnology Unit,
Institute of Public Enterprise,
O U Campus, Hyderabad-500 007

Dr Suman Sahai

Gene Campaign

J-235/A Sainik Farms,
Khanpur,
New Delhi-110 062

Mr Theo van de Sande

Project Officer

Research and Communication Division,
Cultural Cooperation, Education and
Research Department,
Ministry of Foreign Affairs,
Bezuidenhoutseweg 67,
2594 AC The Hague,
The Netherlands.

Chairpersons

Prof E Hari Babu

Department of Sociology,
University of Hyderabad,
Central University PO,
Hyderabad-500 046

Dr K K Sharma

Senior Scientist - Cell Biology

Genetic Transformation Laboratory,
I C R I S A T,
PATANCHERU-502 324

Valedictory Address

Dr Mangala Rai

*Deputy Director General
(Crop Sciences)*

Indian Council of Agricultural Research,
Krishi Bhavan,
New Delhi-110 001

INVITED PARTICIPANTS

Biotechnology Programe Committee Members

Dr G Gopal Reddy

Secretary

Sri Aurobindo Institute of
Rural Development,
Gaddipally-508 201

Dr V P Gupta

Advisor, Department of Biotechnology

Goverment of India,
Block 2, 7th Floor,
CGO Complex, Lodi Road,
New Delhi-110 003

Dr D M Hegde

Project Director
Directorate of Oilseeds Research,
Rajendranagar,
Hyderabad-500 030

Dr K Jayalakshmi

Deputy Director
National Institute for
Rural Development,
Rajendranagar,
Hyderabad-500 030

Smt C S Rama Lakshmi, IFS

Commissoner
Women Empowerment &
Self Employment,
Government of Andhra Pradesh,
Insurance Building, Tilak Road,
Hyderabad-500 001

Dr P Sreeramulu

Additional Director (Production)
Directorate of Animal Husbandry,
Government of Andhra Pradesh,
Shantinagar,
Hyderabad-500 028

Mr Stephen Livera

Executive Director
Society for Development of
Drought Prone Area,
42-189/1, Vengal Rao Colony,
Wanaparthy-509 103

Dr Krishna Ashrit

*Former Director of
Animal Husbandry*
H.No.4-1-150/4/2,
Venkateswara Colony,
Saroornagar, Hyderabad-500 035.

Dr G M Reddy

INSA Senior Scientist
C/o Centre for Plant Molecular Biology,
Osmania University,
Hyderabad-500 007

Others

Dr Ramesh K Agarwal
Centre for Cellular and
Molecular Biology,
Uppal Road, Tarnaka,
Hyderabad-500 007

Prof Aparna Datta Gupta
Department of Animal Sciences,
University of Hyderabad,
Central University PO,
Hyderabad-500 046

Dr B S Bajaj

Chairman
All India Biotech Association,
Southern Chapter,
401, Lakshmi Nivas,
Greenlands, Ameerpet,
Hyderabad-500 016.

Dr S M Balachandran

Senior Scientist
Directorate of Rice Research,
Rajendranagar,
Hyderabad-500 030

Mr D Balakrishna

Scientist
National Research Centre for Sorghum,
Rajendranagar,
Hyderabad-500 030

Dr V Dashavantha Reddy

Principal Coordinator
Centre for Plant Molecular Biology,
Osmania University,
Hyderabad-500 007

Dr V Dinesh Kumar

Senior Scientist
Directorate of Oilseeds Research,
Rajendranagar,
Hyderabad-500 030

Dr N P Eswara Reddy

Associate Professor
Department of Plant Pathology,
S V Agricultural College,
ANGRAU,
Tirupati 517 502

Prof Gopal Reddy

Department of Microbiology,
Osmania University,
Hyderabad-500 007

Dr G Harinarayana

Director
Ganga Kaveri Seeds Pvt. Ltd.,
Suit No.1406/1407,
Babukhan Estate, Basheerbagh,
Hyderabad-500 001

Ms Inge Dijkslag

Assistant Coordinator
TMBT Programme,
Wageningen UR,
Hollandseweg 1,
6706 KN Wageningen,
The Netherlands.

Dr N Jyothi Lakshmi

Scientist
Central Research Institute of
Dryland Agriculture,
Santoshnagar,
Hyderabad-500 059

Prof P B Kavi Kishore

Department of Genetics,
Osmania University,
Hyderabad-500 007

Dr P B Keerti

School of Life Sciences,
University of Hyderabad,
Central University PO,
Hyderabad-500 046.

Prof V V Lakshmi

Head, Department of Microbiology
Sri Padmavathi Mahila
Viswavidyalayam,
Tirupathi-517 502

Ms Thummun Lakshmi

Ph.D.Student
Department of Sociology,
University of Hyderabad,
Central University PO,
Hyderabad-500 046

Dr M Maheswari

Scientist
Central Research Institute for
Dryland Agriculture,
Santhoshnagar,
Hyderabad-500 059

Dr P Maruthi Mohan

Professor & Head Coordinator
UGC-SAP
Department of Biochemistry,
Osmania University,
Hyderabad-500 007

Dr B Prathibha Devi

Department of Botany,
Osmania University,
Hyderabad-500 007

Dr P Raghava Reddy

Principal Scientist (Rice) &
Head
Agricultural Research Station,
Acharya N G Ranga Agricultural
University,
Maruteru-534 122

Prof K V Raman

H.No.1-2-594/B, Gagan Mahal Colony,
Lower Tank Bund,
Hyderabad-500 029

Prof T Ramana

Coordinator, Biotechnology Unit
College of Science & Technology,
Andhra University,
Visakhapatnam-530 003

Dr S V Rao

Scientist
National Research Centre for Sorghum,
Rajendranagar,
Hyderabad-500 030

Dr P V Reddy
Regional Agricultural Research Station,
A N G R A U,
Tirupati-517 502

Dr K R S Sambhasiva Rao
Associate Professor,
Centre for Biotechnology &
Department of Zoology
Nagarjuna University.
Guntur-522 510

Dr N R Shobha Rani
Principal Scientist & Head
Plant Breeding and Genetics,
Directorate of Rice Research,
Rajendranagar,
Hyderabad-500 030

Dr S Sivaramakrishnan
Special Officer, Biotechnology
A N G R A U,
Rajendranagar,
Hyderabad-500 030

Dr S K Soam
Senior Scientist
National Academy of
Agricultural Research Management,
Rajendranagar,
Hyderabad-500 030

Dr P Srinivasula Reddy
Associate Professor & Head
Department of Biotechnology,
Sri Venkateswara University,
Tirupathi-517 502

Prof C Subrahmanyam
Department of Biochemistry,
Osmania University,
Hyderabad-500007

Dr C Sudhakar
Associate Professor
Department of Botany,
Sri Krishnadevaraya University,
Ananthapur-515 003

Dr M R Sundaram
*Scientist, Directorate of
Rice Research*
Rajendranagar,
Hyderabad-500 030

Dr K Ulaganathan
Centre for Plant Molecular Biology
Osmania University,
Hyderabad-500 007

Dr K Uma Devi
Associate Professor
Department of Botany,
Andhra University,
Visakhapatnam-530 003

Dr P Uma Maheswari Devi
Assistant Professor
Department of Applied Microbiology,
Sri Padmavathi Mahila Viswavidyalayam,
Tirupati-517 502

Dr S Vasanthi
Scientific Officer
National Institute of Nutrition,
Jamai Osmania,
Hyderabad-500 007

Sri S Venkat Reddy
Hon' Secretary & Treasurer
Seedsman Association,
Room No.50, 3rd Floor,
Abids Shopping Centre,
Opp. Emrold Hotel,
Hyderabad-500 001

Dr K B R S Visarada
Senior Scientist
National Research Centre for Sorghum,
Rajendranagar,
Hyderabad-500 030

Dr S K Yadav
Senior Scientist
Central Research Institute for
Dryland Agriculture,
Santoshnagar,
Hyderabad-500 05

List of Abbreviations

ABRE	:	ABA Responsive Element
AFLP	:	Amplified Fragment Length Polymorphism
AIA	:	Advanced Informed Agreement
APNLBP	:	Andhra Pradesh Netherlands Biotechnology Programme
AP-PCR	:	Arbitrary Primed PCR
ASAP	:	Allele Specific Associated Primer
ASI	:	Anthesis to Silking
BARC	:	Bhabha Atomic Research Centre
BCIL	:	Biotechnology Consortium of India Limited
BLB	:	Bacterial Leaf Blight
BNF	:	Biological Nitrogen Fixation
BNL	:	Brookhaven National Laboratory
BPC	:	Biotechnology Programme Committee
BPH	:	Brown Plant Hopper
BSA	:	Bulked Segregant Analysis
BT	:	Biotechnology
BTU	:	Biotechnology Unit
CAC	:	Codex Alimentarius Commission
CAPS	:	Cleaved Amplified Polymorphic Sequence
CBD	:	Convention on Biological Diversity
CCMB	:	Centre for Cellular and Molecular Biology
CIM	:	Composite Interval Mapping
DAF	:	DNA Amplification Fingerprinting
DAMD	:	Directed Amplification of Minisatellite-Region DNA
DBT	:	Department of Biotechnology
DNA	:	De-oxy Ribonucleic Acid
DRE	:	Dehydration Responsive Element
DRR	:	Directorate of Rice Research
EPP	:	Ears Per Plant

EU	:	European Union
FAO	:	Food and Agriculture Organization
FDA	:	Food and Drug Administration
FFLW	:	Female Flowering
FI	:	Fluorescence Intensity
FRLHT	:	Foundation for Revitalization of Local Health Tradition
GDP	:	Gross Domestic Product
GE	:	Genotype x Environment
GEAC	:	Genetic Engineering Approval Committee
GM	:	Genetically Modified
GOT	:	Grow-Out Test
GRWT	:	Grain yield
HBV	:	Hepatitis B Virus
IBSC	:	Institutional Biosafety Committee
IBU	:	Interactive Bottom Up
ICAR	:	Indian Council for Agricultural Research
ICMR	:	Indian Council for Medical Research
IMWIC	:	International Maize and Wheat Improvement Centre
INGER	:	International Network for Genetic Evaluation of Rice
IPM	:	Integrated Pest Management
IRGSP	:	International Rice Genome Sequencing Project
IRRI	:	International Rice Research Institute
ISSR	:	Inter Simple Sequence Repeat
IT	:	Information Technology
KARI	:	Kenya Agricultural Research Institute
LMOs	:	Living Modified Organisms
LOD	:	Log of Odds
LPA	:	Lysophosphatidic Acid
LR	:	Likelihood Ratio
MAG	:	Monoacylglycerol
MAS	:	Marker Assisted Selection
MFLW	:	Male Flowering
MPCA	:	Medicinal Plant Conservation Areas
MTA	:	Material Transfer Agreement
NGO	:	Non Governmental Organisation

OECD	:	Organization for Economic Cooperation and Development
PCR	:	Polymerase Chain Reaction
PGS	:	Plant Genetic Systems
PLHT	:	Plant Height
PPFW	:	Pre-Project Formulation Workshops
PSCL	:	Proagro Seed Co. Ltd.
PTD	:	Participatory Technology Development
QTL	:	Quantitative Trait Loci
R & D	:	Research and Development
RAMPO	:	Random Amplified Microsatellite Polymorphism
RAPD	:	Random Amplified Polymorphic DNA
RCGM	:	Review Committee for Genetic Modifications
RDAC	:	Recombinant DNA Advisory Committee
RFLP	:	Restriction Fragment Length Polymorphism
RILs	:	Recombinant Inbred Lines
SAMPL	:	Selective Amplification of Microsatellite Polymorphic Loci
SCAR	:	Sequence Characterized Amplified Region
SPS	:	Sanitary and Phytosanitary Measures
SSCP	:	Single Strand Conformation Polymorphism
SSR	:	Simple Sequence Repeat
STMS	:	Sequence Tagged Microsatellite Sites
STSs	:	Sequence Tagged Sites
TAG	:	Triacylglycerol
TBT	:	Technical Barriers to Trade
TGMS	:	Thermosensitive Genetic Male Sterility
UGC	:	University Grants Commission
UMC	:	University of Missouri
UNDP	:	United Nations Development Programme
USA	:	United States of America
VNTR	:	Variable Number of Tandem Repeats
WHO	:	World Health Organization
WTO	:	World Trade Organization